# THE UNOFFICIAL

# IEEE BRAINBUSTER GAMEBOOK

Mental Workouts
for the Technically Inclined

Compiled by

**Donald R. Mack**

Ph.D., P.E.,
Life Senior Member, IEEE

IEEE
Press

The Institute of Electrical and Electronics Engineers, Inc., New York

William Perkins, *Editor in Chief*
Dudley R. Kay, *Executive Editor*
Carrie Briggs, *Administrative Assistant*
Denise Gannon, *Production Supervisor*
Marybeth Hunter, *Marketing Manager*


This book may be purchased at a discount from the publisher when ordered in bulk quantities. For more information contact:

IEEE PRESS Marketing
Attn: Special Sales
PO Box 1331
445 Hoes Lane
Piscataway, NJ 08855-1331
Fax: (908) 981-8062


©1992 by the Institute of Electrical and Electronics Engineers, Inc.
345 East 47th Street, New York, NY 10017-2394

All rights reserved. No part of this book may be reproduced in any form, nor may it be stored in a retrieval system or transmitted in any form, without written permission from the publisher.

Printed in the United States of America

10  9  8  7  6  5  4  3  2


ISBN 0-7803-0423-3

IEEE Order Number: PP0318-6


**Library of Congress Cataloging-in-Publication Data**

Mack, Donald R.
 The unofficial IEEE brainbuster gamebook: mental workouts for the technically inclined / Donald R. Mack.
   p.  cm.
  ISBN 0-7803-0423-3
  1. Mathematical recreations.   I. Institute of Electrical and Electronics Engineers.   II. Title
  QA95.M24      1992
  793.7'4—dc20                                             92-4724
                                                              CIP

# Contents

Preface — v

1. Problems Requiring Only Logic — 1

2. Problems Requiring Some Mathematics — 23

3. Little Engineering Problems — 39

4. Solutions to the Problems — 59

5. Acknowledgments — 135

# Preface

The problems in this book have appeared in the Gamesman section of the *Potentials* magazine, a quarterly published for engineering students by the Institute of Electrical and Electronics Engineers. Most of the problems were submitted by student readers, and some by experienced engineers who enjoy logical, mathematical, and technical puzzles. If you were interested enough to pick up the book, you are probably already familiar with some of the mental challenges on its pages. A few of the problems were probably familiar to Pythagoras. The challenge, not the age, of each problem was the criterion for inclusion. The reader who first submitted it to *Potentials* gets the credit in the list in Chapter 5.

# Problems Requiring Only Logic

1.1. Lined up in the dusty stacks of a college library are the volumes of IEEE *Potentials*. Each volume consists of four issues, each 5 mm thick. The front and back covers of each volume are 4 mm thick. A bookworm begins at the first page of volume 1, issue 1 and eats in a straight line until she reaches the last page of volume 3, issue 4. How far does the bookworm eat?

1.2. If you have a five-liter and a three-liter bottle and plenty of water, how can you get four liters of water in the five-liter bottle?

1.3. You encounter three people who know each other. One always tells the truth, one lies all the time, and one gives random answers. How can you tell, by asking only three questions directed to only one person at a time, which is which?

# Problems Requiring Only Logic

1.4. You have 10 sets of 10 coins each. Each of the coins weighs 10 grams, except for the coins in one set, which weigh 9 grams each. All 100 coins look alike. How can you identify the set of 9-gram coins with only one weight measurement?

1.5. In Bill's sock drawer are eight pairs of white socks and six pairs of red socks. (He's a snappy dresser.) How many socks does Bill have to pick at random to be sure he has a matched pair?

1.6. This problem is for chess players. Place a knight in one of the corner squares of a chessboard. Using his move, land in all 64 squares without reaching the same square twice.

1.7. Connect the nine points below with four straight lines, without lifting the pencil from the paper.

. . .

. . .

. . .

1.8. This one is easy, too, if you don't make an unstated assumption. Make four equilateral triangles with these six sticks of equal length:

1.9. Three engineers rented a hotel room that was

supposed to cost $40 for the night (a long time ago). The desk clerk mistakenly charged them $15 each, payable in advance. Later he realized he had overcharged them, but couldn't figure out how to divide the five-dollar refund among the three engineers. So he pocketed two dollars and returned one dollar to each of the engineers. The engineers ended up paying $14 each, or $42. That, plus the clerk's two dollars, added up to $44. What happened to the other dollar?

1.10. At a political convention there are 100 politicians. Each one is either crooked or honest. At least one is honest. Given any two of the politicians, at least one is crooked. How many are honest and how many are crooked?

1.11. The inhabitants of North Strangeland are of either type A or type B. Type A people can ask only questions whose correct answer is *"yes."* Those of type B can ask only questions whose correct answer is *"no."* George was overheard to ask, "Are Nancy and I both type B?" Which type is Nancy?

1.12. In South Strangeland, sane humans and insane vampires make only true statements. Insane humans and sane vampires make only false statements. Two sisters, Anna and Betsy, live there. We know that one is a human and one is a vampire, but we know nothing about the sanity of either. Anna says, "We are both insane." Betsy replies, "That's not true!" Which sister is the vampire?

1.13. I have two coins that total 55 cents. One of them is not a nickel. What are the two coins?

1.14. An archeologist walking along the shore of the Mediterranean Sea finds an old Roman coin. On one side is the face of Julius Caesar and the date 44 BC. On the other side is an olive tree. The archeologist says, "This coin is counterfeit." How does she know?

1.15. There are nine coins in a bag. One of them is counterfeit. A real coin weighs one gram and a counterfeit one weighs 0.9 gm. You have a balance. How can you identify the counterfeit coin with two weighings?

1.16. Two friends, George and Harry, were born in May, one in 1964 and the other a year later. Each has an antique 12-hour clock. One clock loses 10 seconds an hour and the other gains 10 seconds an hour. On a day in January the two friends set both clocks right at exactly 12 noon. "Do your realize," says George, "that the clocks will drift apart and won't be together again until . . . good grief, your 23rd birthday!" How long will it take for the two clocks to come together again? Which friend is the older, George or Harry?

1.17. The dean of engineering had three 4.0 seniors, one of whom was blind. To decide which one would be the valedictorian, he devised a tie breaker. He explained to the students that he had collected five hats: three white and two red. Then when their eyes were closed he put one on each student's head. "Now open your eyes," he said, "and tell me the color of the hat on your head." After a while one of the students who could see said, "I can't be sure." A minute later the second said, "Neither can I." Then "Aha!" cried the blind student. "My hat must be white!" Please explain how the blind student accomplished this impressive reasoning.

1.18. In a remote village in Asia everyone either tells the truth all the time or lies all the time. An English-speaking traveler encounters two of the villagers and asks Mr. A, "Are you a truth teller or a liar?" Unfortunately Mr. A speaks no English, so Mr. B translates the question and the answer, and replies to the traveler, "Mr. A says he is a liar." Now, which is Mr. B, a truth teller or a liar?

1.19. Two trains, each moving at 50 km/hr, were approaching each other on the same track. When they were 100 km apart, a bee on the front of one train started flying toward the other train at a steady ground speed of 60 km/hr. When it reached the other train, it immediately started back toward the first train. It continued to fly back and forth until the trains collided. How far did it fly? Incidentally, the bee escaped.

1.20. An oarsman leaves his boathouse on the river and rows upstream at a steady rate. After 2 km he passes a log floating down the river. He continues on for another hour, and then turns around and rows back downstream. He overtakes the log just as he reaches the boathouse. What is the flow velocity of the river?

1.21. I know two lumberjacks. The little lumberjack is the son of the big lumberjack, but the big lumberjack is not the little lumberjack's father. Who is the big lumberjack?

1.22. A monk started up a hill one morning and reached the top at sunset. He stayed overnight in the temple there, started back down the hill the next morning by the same path, and arrived at the bottom at sunset. Was there a place on the path that he passed at exactly the same time of day going both ways?

1.23. This one is difficult. A married couple invited other married couples to a party. Including the host and hostess, there were $n$ couples. As people arrived, they shook hands with only the people they had not met before. Eack person already knew his or her own partner, of course. After everyone had arrived, the host asked each person, including the hostess, how many times he or she had shaken hands. To everyone's surprise, each person responded with a number different from everyone else's. How many times did the hostess shake hands? For extra credit, calculate how many time the host shook hands.

1.24. Draw the next figure without lifting the pencil off the paper or retracing any line.

# Problems Requiring Only Logic

1.25. The people who live on the east side of the city of Loopbow always tell the truth, and the people who live on the west side always lie. During the day the people mingle on both sides. What one question can you ask a resident picked at random to determine which side of the city you are in?

1.26. A traveler is driving to Millinocket. He comes to a fork in the road and sees a farmer. The farmer is from one of two families, one who are truth tellers, and one who always lie. What question can the traveler, knowing this, ask the farmer to find out which is the road to Millinocket?

1.27. Please find the words whose initials are on the right side of each equation below. The first answer is shown.

1. 26 = L of the A  (Letters of the alphabet)

2. 7 = W of the A W

3. 1,001 = A N

4. 12 = S of the Z

5. 54 = C in a D (including the J)

6. 9 = P in the S S

7. 88 = K on a P

8. 13 = S on the A F

9. 32 = D F at which W F

10. 18 = H on a G C

11. 90 = D in a R A

12. 200 = D for P G in M

13. 8 = S on a S S

14. 3 = B M (S H T R)

15. 4 = Q in a G

16. 24 = H in a D

17. 1 = W on a U

18. 57 = H V

19. 11 = P on a F T

20. 1,000 = W that a P is W

21. 29 = D in F in a L Y

22. 64 = S on a C

23. 40 = D and N of the G F

## Problems Requiring Only Logic

1.28. How can you distribute 1023 coins in 10 bags so that you can provide any number of coins without opening a bag? Start with the low numbers. You need a bag with just one coin in it, and one with just two, but you don't need a bag with just three, etc.

1.29. A strange monetary system has been proposed that will use International Monetary Units, or IMUs, instead of dollars or yen. All prices will be in integral numbers of IMUs. The money will consist of paper bills in 10 denominations that will allow you to make any purchase from 1 to 1023 IMUs without having to use more than one bill of each denomination. What should the denominations of the bills be? For extra credit, explain why showing the value of each bill and the prices of purchases in *binary* numbers would simplify paying for a purchase.

1.30. Please fill in the blank. Pedro is the son of Juan. Juan is the _____ of Pedro's father.

1.31. An engineer got a flat tire. After removing the four lug nuts from the wheel, he accidentally dropped all four down a sewer drain. He knew that the wheel would stay on with only three lug nuts. He thought of stealing three nuts from one of the cars parked nearby, but that would be dishonest, and the stolen nuts probably wouldn't fit anyway. Can you suggest a practical solution to his problem?

1.32. You have six bottles of pills. All of the pills look alike and each weighs one gram, except for the pills in one of the bottles that weigh two grams each. How can you determine, with *one weighing*, which bottle contains the two-gram pills? You don't need a balance. A spring scale will do.

1.33. On one side of a river are the king (K), the queen (Q), the prince (P), the princess (PR), a guard (G), and the guard's wife (W). They have to cross the river in a two-passenger rowboat, so only one or two can cross at a time. They are all able to row the boat. The crossings are subject to two ridiculous conditions:

- None of the royal family may travel with a commoner.

- No man may be on the same side of the river with a woman other than his wife, even if he is in the boat and she is on the shore, or vice-versa, unless the woman's husband is present.

What is the sequence of trips that will get all six people across the river?

1.34. What are the next two letters of this sequence?

OTTFFSS _ _

1.35. Sketch the next two symbols in this sequence:

## Problems Requiring Only Logic

1.36. A gold miner lives at point A, six km north of the river on the map below. His mine is at point B, 15 km from his cabin and three km north of the river. He can't go directly to work because his burro has to stop at the river on the way, for a drink. At what point on the river should he water the burro, to minimize the length of the trip?

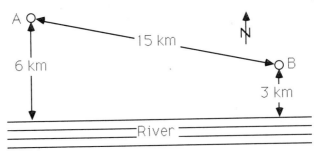

1.37. I have a chain of 23 paper clips. If I remove two of them, I can make a chain of any length from one to 23 out of the five pieces. Which two should I remove?

1.38. A fastidious fellow wears a clean shirt every day. Every Friday morning he drops off the week's laundry and picks up the previous week's laundry. How many shirts must he have?

1.39. A student purchased his books for the spring quarter. He noticed that all but two had blue covers, all but two had red covers, and all but two had green covers. How many books did he have?

# Problems Requiring Only Logic

1.40. On the shelf are three cans, labeled "NUTS," "BOLTS," and "NUTS AND BOLTS." We know that the labels are mixed, so that none is on the right can. How many samples must we take (without looking into the cans) to establish the correct labeling?

1.41. What is the largest amount of change in U.S. coins you can have in your pocket and still not have change for a dollar bill?

1.42. There are four students of different weights, heights, and ages. The freshman, who is the youngest, is shorter than the sophomore, who is the heaviest. The sophomore is younger than the junior, who is the tallest. How does the senior compare with the others if no one occupies the same rank in any two of the categories?

1.43. A creative teacher arranged the chairs for her students in five rows, with four chairs per row. But there were only 10 students. How did she arrange the chairs?

1.44. How can you cut a round pizza into eight pieces with three cuts? Don't try any tricks like retracing a cut, folding the pizza, or removing it from the pan.

1.45. Next are four views of the same cube. Which face is opposite the C?

# Problems Requiring Only Logic

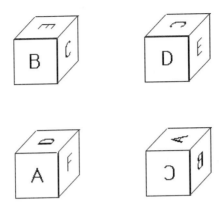

1.46. The square below has an area of 64 square units. When its parts have been reassembled to form the rectangle, the area is 65 square units. What's wrong?

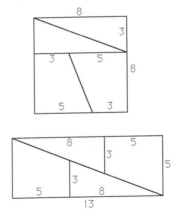

1.47. Eight cardboard squares of equal size are stacked as shown on the next page. Please number them, starting with the one on the bottom.

**Problems Requiring Only Logic**

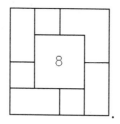

See if you can decode the following puzzles. Here is an example:

    MAN                    Solution: Man overboard
    BOARD

1.48. ALL world

1.49. SYMPHON

1.50. GROUND
      FEET
      FEET
      FEET
      FEET
      FEET

1.51.     J
    YOU U ME
       S
       T

1.52. STAND
     I

1.53. WEAR
     LONG

## Problems Requiring Only Logic

1.54. CYCLE
CYCLE
CYCLE

1.55. DICE
DICE

1.56. EILN PU

1.57. $\dfrac{\text{NaCl.H}_2\text{O}\ \ \ \ \ }{\text{CCCCCCC}}$
NaCl.H$_2$O

1.58. WHEATHER

1.59. LE
    VEL

1.60. 1 3 5 7 9
WHELMING

1.61. (CAPITALISM)$^{HE}$

1.62.

1.63.

1.64. ECNALG

1.65. <u>　0　</u>
B.S.
M.S.
Ph.D.

1.66. ZERO
DEGREE
DEGREE
DEGREE

1.67. ONCE
LIGHTLY

1.68. NOON GOOD

1.69. STROKES
strokes
*strokes*

1.70. AGE BEAUTY

1.71. MAN
campus

1.72. GETTINGITALL

# Problems Requiring Only Logic

1.73. D
DEER
 E
 R

1.74. R|E|A|D|I|N|G

1.75. T
O
U
C
H

1.76. HOU
 SE

1.77. John's father has three children. One is Richard, who lives in Chicago. Another is Margaret, who lives in Vancouver. Who is the third?

1.78. Susie ChemE, conducting an experiment in the chemistry laboratory, poured 54 cc of water into a beaker from a graduated cylinder. Then she accidentally dropped the cylinder and broke it. She had meant to pour only 50 cc of water into the beaker. She found three large identical test tubes, and with eight pourings back and forth between the test tubes and the beaker she was able to end up with 50 cc in the beaker. Can you duplicate her feat? Susie had a steady hand and could pour equal amounts of water into two or three test tubes. Consider such an operation as one pouring.

1.79. Ms. ChemE hasn't replaced the graduated

cylinder yet. Now she has three bottles that hold 8, 5, and 3 cc. The 8-cc bottle is full of a solution that she must divide into two equal parts. She can do it with seven pourings. Can you?

1.80. Put six glasses upright on a table. Invert five of them. Now invert another combination of five. How many such moves do you need to get all of the glasses upside down?

1.81. Draw the figure below with at most 22 strokes, without lifting the pencil off the paper or retracing a line.

1.82. Following is the Monty Hall problem, named for the host of the television game show "Let's Make a Deal." Thousands of intelligent people, including professional mathematicians, have solved it incorrectly.

On a game show are three doors. The host says, "Behind one door is a Miata, and behind the other two are goats. I know which door is which. Please point to a door. Then I'll open one of the other two doors to show a goat. After you've seen the goat, you may switch your choice to the third door, or keep your original choice. You'll win whatever is behind the door of your final choice." Should you switch, or stay with your original choice, or does it matter?

1.83. If you understand the solution to the Monty Hall problem, try this extension. Now there are $n$ doors. Behind one is a Miata and behind the other $n-1$ are goats. The host lets you choose a door, then opens $n-2$ of the others, revealing $n-2$ goats, and again gives you the option of switching your choice. What is your probability of winning the Miata if you stay with your original choice, or switch to the last door?

1.84. Five open triangles (triangles with no lines crossing them) can be formed with five straight lines, as shown below. What is the maximum number of open triangles that can be formed with seven straight lines?

1.85. If the puzzle you solved before you solved the puzzle you solved after you solved the puzzle you solved before you solved this one, was harder than the puzzle you solved after you solved the puzzle you solved before you solved this one, was the puzzle you solved before you solved this one harder than this one?

1.86. According to legend, one day when the mathematician K. F. Gauss was a schoolboy, his teacher wanted to leave the classroom for a while. To keep the students busy he asked them to add all the integers from one to 100. Before he was out of the room, Gauss had the answer. Can you make the calculation in less time than it takes a teacher to leave the room?

1.87. This problem is being given to eighth graders in Vermont, in these words: How many different squares of all sizes are on a checkerboard? *Hint:* Organize your listing. Look for patterns.

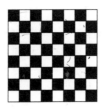

# Notes & Calculations

# Problems Requiring Some Mathematics

2.1. Which is larger, $e^\pi$ or $\pi^e$? See if you can find out analytically, without the aid of a calculator or computer.

2.2. An engineer entered the north end of a tunnel of length $L$. After walking the distance $L/4$ into the tunnel, he noticed a car approaching the north entrance at 40 miles per hour. The engineer knew his own speed, and calculated that no matter which end of the tunnel he ran to, he would arrive there at the same time as the car. What was his top speed? *Hint*: he might do better as a professional athlete than as an engineer.

2.3. What is the missing number in the series below? *Hint*: they are all the same number in different bases.

10, 11, 12, 13, 14, 15, 16, 17, 20, 22, 24, ___, 100, 121, 10000

**2.4.** When an airplane recently flew from Los Angeles to San Francisco with a moderate headwind, its average speed was 380 mph. On the return trip, with the wind, its average speed was 420 mph. Was the average speed for the round trip 400 mph? If not, what was it?

**2.5.** Which is larger, $\sqrt{17} + \sqrt{10}$ or $\sqrt{53}$? No calculators or tables, please.

**2.6.** What is the smallest number which, if you move its most significant (left-most) digit all the way to the right to make it the least significant digit, becomes half of the original number? Don't spend a lot of time trying to solve this one. Just look at the answer.

**2.7.** In order to solve this and the next six problems, you have to know some probability theory. I know a couple that has five daughters and no sons. What is the probability that five consecutive children will all be girls?

**2.8.** A missile's guidance system contains 100 transistors, each of which is 99.9 percent reliable. The failure of any one of them will cause the guidance system to fail. How reliable is the guidance system?

**2.9.** Craps is played with two dice. The thrower wins on the first throw if he rolls a 7 or 11, and loses if he rolls a 2, 3, or 12. If he makes any other point, he throws again until he makes that point again, in which case he wins. If he rolls a 7 before making his point, he loses. In a crap game is it wiser to take the dice or to bet against the thrower?

2.10. How many people do you need to have in a group to be 50 percent sure that at least two of them have the same birthday? Don't worry about birthdays on February 29.

2.11. A traffic light is red for 30 seconds and green for 30 seconds. (It has no amber.) What is the expected wait at the light?

2.12. You and your friend are going to take turns rolling a die or flipping a coin, with a probability $p$ of winning on any try. You'll keep it up until one of you wins. How important is it to go first? Specifically, if you are first, what is your probability of winning and what is your friend's probability?

2.13. What is the probability that if you cut a straight stick at two arbitrary places, you can make a triangle with the three pieces?

2.14. The host at a party said, "I have three daughters. The product of their ages, all whole numbers, is 72. The sum of their ages is my house number. Can you tell me how old they are?" The guest, knowing the house number, did some calculation on his portable PC, and then realized that he needed more information. The host added, "The oldest girl likes puzzles." "That's what I needed," said the guest. "Now I know their ages." Do you?

2.15. Here's a dilemma. Start with the equality

$$x^2 = x + x + x + \ldots \text{ (}x\text{ times)}.$$

Differentiate both sides to get

$$2x = 1 + 1 + 1 + \ldots \text{ (}x\text{ times)}$$
$$= x.$$

What's wrong?

2.16. What is the next number in this sequence?

$$1, 2, 6, 12, 60, 60, 420, 840, \ldots$$

2.17. You're negotiating to do a job that will take all of January, 31 days. The employer offers to pay you one cent for the first day, two cents for the second, and so on, doubling the amount each day. "Alternatively," he says, "I'll pay you a flat one hundred million dollars for the job." Which payment will you choose?

2.18. Fill in the nine squares below with the numbers 1 to 9, using only one of each, so that the sum of any three numbers obtained by adding them horizontally, vertically, or diagonally is 15.

## Problems Requiring Some Mathematics

2.19. Write the code that will swap the values of two variables A and B without using a third variable.

2.20. A box contains one ball. We don't know whether the ball is black or white, so the probability of either is 1/2. An engineer drops in a black ball. Then he reaches in and removes a ball, which turns out to be black. "Aha," he says, "Now the probability that the other ball is black is 2/3." Is he right, or is the probability still 1/2?

2.21. Here's an interesting gambling game. You pay the house a fee for the privilege of flipping a coin. The house will pay you a dollar for each flip it takes you to get the first head. What should the fee be to make it a fair game? You'll need this item:

$$\sum_{n=1}^{\infty} n/2^n = 2 \ .$$

2.22. Suppose that the house will pay you $2^n$ dollars if it takes $n$ flips to get the first head. Now what is a fair flipper's fee?

2.23. The three barrels shown on the next page are tied together. The diameter of each barrel is one meter. How long is the rope?

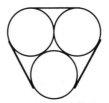

2.24. Here's a more difficult geometry problem that proves a strange fact. Choose a solid sphere, shown below, that is at least 6 cm in diameter. Drill a hole through its center, choosing the diameter of the hole so that its length is 6 cm. The strange fact is that no matter how big the sphere is, the volume of material remaining is the same. What is that volume?

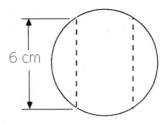

2.25. An airplane having a cruising speed of 200 km/hr flies from one airport to another 200 km away and then returns. How long does the round trip take if

1. There is no wind?
2. There is a constant 40 km/hr headwind going and tailwind returning?
3. There is a constant 40 km/hr crosswind?

2.26. A cereal manufacturer puts a coupon, numbered 1 to 5, in each box. When you collect a set of all five numbers, you can send in for a "valuable free gift." On the average, how many boxes do customers have to buy to win a prize?

2.27. A new drag race has been proposed that probably won't catch on. Four racers start at the corners of a one-kilometer square, shown below. They start at the same time and accelerate at the same rate, each aiming at the car that started at the corner on his left. What is the length of the spiral path each car takes before they all collide?

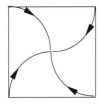

2.28. Bill and Mary run two laps around a track. Mary runs at the same speed all the way. Bill lets her get ahead by running for one lap at her speed less $x$, and then shows off by running the second lap at her speed plus $x$. Who wins the race?

2.29. The axes of two circular cylinders of unit radius intersect at right angles. What is the volume common to both cylinders?

2.30. Two men are walking at different steady paces upstream along the bank of a river. A ship moving downstream at constant speed takes ten seconds to pass the first man. Ten minutes later it reaches the second man, and takes nine seconds to pass him. Starting then, how long will it take for the two men to meet?

2.31. Please write down the next row at the bottom of this table:

        1
        1 1
        2 1
        1 2 1 1
        1 1 1 2 2 1
        3 1 2 2 1 1
        1 3 1 1 2 2 2 1

Why will only the numbers 1, 2, and 3 ever appear in the table?

2.32. How can you obtain 3 using only zeros and common mathematical symbols?

2.33. The figure below shows two concentric circles and a chord of the outer circle tangent to the inner circle. How can you calculate the annular area between the circles if all you know is the length of the chord?

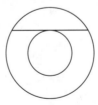

2.34. James beat Kidd in a set of tennis, 6 to 3. Five of the games were won by the receiver. Who served first?

## Problems Requiring Some Mathematics

2.35. Michael tried to impress Dorothea by taking her to an expensive restaurant. To try to save money, he ordered only one dessert. It came in a parabolic glass, shown below. He asked Dorothea to eat only half and leave the rest for him. If the original depth of the dessert was $h$, what is the depth $x$ of the amount Dorothea should leave for Michael? As you can imagine, that was Michael's last date with Dorothea.

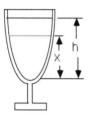

2.36. If
$$y = \sqrt{\sin x + \sqrt{\sin x + \sqrt{\sin x + \ldots}}},$$
what is $dy/dx$?

2.37. Calculate the numerical value of the infinite progression
$$z = \sqrt{2 + \sqrt{2 + \sqrt{2 + \sqrt{2 + \ldots}}}}.$$

2.38. If $a$ through $z$ are 26 constants, what is the numerical value of
$$(n - a)(n - b)\cdots(n - z) \ ?$$

2.39. Show that $j^j = 0.208$.

2.40. Here are four interesting number games.

A. Write 100 using five 5's.

B. What is the smallest integer that can be written with two digits?

C. Write a number equal to 1 using all 10 digits.

D. What is the highest number that can be written with four 1's?

2.41. Construct a Fibonacci series by starting with any two numbers and then adding more, each being the sum of the preceding two. For example, -2, 3, 1, 4, 5, 9, 14, 23, 37, 60, 97, 157. After you've added about 10 terms, the quotient of the last two is 1.618. Can you explain why?

2.42. If you multiply a column vector **a** by a row vector **b**:

$$\begin{bmatrix} a_1 \\ a_2 \\ \cdot \\ \cdot \\ a_n \end{bmatrix} [b_1, b_2, \cdots, b_n]$$

you get a square $n \times n$ matrix. I know in advance what the determinant of the matrix is. Do you?

2.43. An indecisive lady plans to build a pool whose bottom, shown next, is made of square tiles of equal size and will consist of a 6 x 6 square of blue tiles surrounded by a single row of white tiles. When they arrive she decides she wants white tiles in the center, surrounded by a single row of blue tiles. She can make the change and use all the tiles by just making the pool rectangular. What will the new dimensions be?

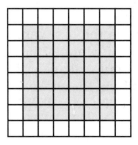

2.44. Starting now, Bill can get to class at 3 PM if he rides his bike at 10 km/hr, and at 1 PM if he rides at 15 km/hr. But the class starts at 2 PM. He should ride at 12.5 km/hr, right?

2.45. Around 1637 the French mathematician Pierre de Fermat wrote a theorem in the margin of a book, that can be stated like this: no combination of nonzero integers $x$, $y$, and $z$ can be found that satisfies the equation

$$x^n + y^n = z^n,$$

if $n$ is greater than 2. He claimed that he had a proof, but the margin was too narrow to include it. For 355 years the best mathematical minds have been unable to prove or disprove this famous "Fermat's last theorem." Is there another Fermat out there? If you can present a proof, you will become famous.

2.46. The World Series is played between the winners of the American and National baseball leagues. The winner of the series is the first team to win four games. Suppose the two teams are evenly matched, so that the probability of either team winning any game is 0.5. Before they start, the probability that team A will win the series is 0.5. Team A wins the first game. Now what is its probability of winning the series? If you can figure that out, calculate the probability that team A will win the series if it wins the second game. What if it wins the third game?

2.47. The solution to this problem is not easy, but it is straightforward. Find a number, which if divided by

    2, leaves a remainder of 1
    3                             2
    4                             3
    5                             4
    6                             5
    7                             6
    8                             7
    9                             8

2.48. In the three boxes shown next are two black balls, two white balls, and one black and one white. You can't see inside. You choose one of the boxes. The probability that you have chosen box 1 is 1/3, right? Now you reach in without looking, and draw out a black ball. Now what is the probability that you have chosen box 1? Is it still 1/3?

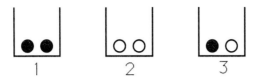

2.49. The figure below shows two islands in a river, connected to the north and south banks and to each other by five drawbridges. Slightly unrealistic conditions are that the probability of each bridge being up is 0.5, and these five probabilities are independent of each other. What is the probability that a car can cross the river at any given time?

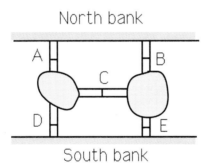

# Notes & Calculations

**Notes & Calculations**

# Little Engineering Problems

3.1. The crew of a tugboat on Puget Sound has found a rectangular box floating in the water with its top horizontal. The top is square, 1 m x 1 m. When the crew lifts the box a little with its crane and then releases it, it bobs up and down in the water with a period of two seconds. The crew wonders if the box can be lifted out of the water. Can you tell how much it weighs? Assume that the density of the water in Puget Sound is 1025 kg/m$^3$.

3.2. Suppose that a construction company digs a round tunnel straight down to the center of the earth and out to the other side, for example from South America to China. If the construction workers overcome the technical difficulties such as keeping the tunnel cool near the middle, evacuating the air from the tunnel, and eliminating friction on the tunnel walls, cylindrical cars can fall freely down the tunnel and come to rest just as they reach the surface on the other end. Before solving the construction problems, solve this one: How long will the trip take? The radius of the earth is $6.371 \times 10^6$ meters. Assume that the density of the earth is uniform.

3.3. How long does it take a low-level satellite to make one trip around the earth? For this calculation assume that the satellite flies at sea level with no wind resistance, so its distance from the center of the earth is the earth's radius, $6.371 \times 10^6$ meters.

3.4. You know the altitude of a geosynchronous satellite. Show how to calculate it.

3.5. If a meteorite drifts slowly into the earth's gravitational field, how fast will it be moving when it reaches the earth's atmosphere, at an altitude of about 200 km?

3.6. What is the ratio of height to base diameter of a cylindrical can that encloses a given volume with a minimum area of metal? Nontechnical question for extra credit: Why don't the cans on grocery store shelves have this ratio?

3.7. At what elevation angle will a gun shoot the greatest distance over level ground?

3.8. Electrical engineers should be able to fill in the four blanks:

10,11,12,13,15,16,18,20,22,24,27,30,33,36,39,43,

47, 51, 56, 62, ___ , ___ , ___ , ___ , 100

3.9. The figure below shows a cylindrical tank filled with water to a depth $d$. At what depth $h$ should a hole be drilled to allow water to squirt the greatest distance from the tank? Leonardo da Vinci solved this problem about 500 years ago, and he didn't know calculus.

The velocity of the water squirting out of the hole is

$$v = c\sqrt{2gh},$$

where $c$ is a coefficient less than 1. It approaches 1 as the efficiency of the hole in converting the potential energy of the water to kinetic energy approaches 100 percent.

If $c = 1$, how far does the water squirt?

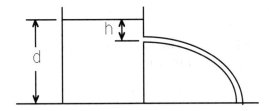

3.10. The right-hand capacitor in the figure below is

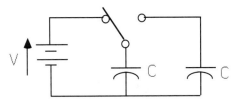

initially uncharged. When the switch is in the left position, the left capacitor has the voltage $V$, the charge $CV$, and the energy $CV^2/2$. When the switch is thrown to the right, the charge redistributes to make the voltage on each capacitor $V/2$. The energy stored in each capacitor is now $CV^2/8$. That accounts for the energy $CV^2/4$. What happened to the other half of the original energy?

3.11. Weights $m_1$ and $m_2$ with masses of 10 and 30 kg hang on a rope around a lightweight pulley with a well-oiled bearing, as shown in the figure below. What is the acceleration of the weights?

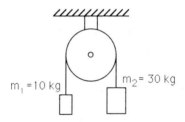

3.12. The figure below shows a horizontal string of length $L$ tied to a pin at its right end and to a

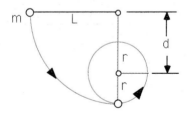

small mass $m$ at its left end. The mass falls through an arc of 90 degrees. Then the string

# Little Engineering Problems

hits a second pin that is the distance *r* above the mass, and starts to swing upward in a circle of radius *r*. If the distance $d = L - r$ is too short, the mass will fall on one side of the pin or the other instead of wrapping around it. What is the minimum distance *d* that will allow the mass to wrap itself around the lower pin?

3.13. The curve shown below, whose formula is

$$y = \frac{1}{1 + x}$$

extends to infinity in the *x* direction. Make a cornucopia by revolving the curve about its *x* axis. I claim that the cornucopia has an infinite inner surface area but encloses a finite volume. I could paint that infinite area by filling the cornucopia with a finite volume of paint and pouring out the excess. Am I right or wrong?

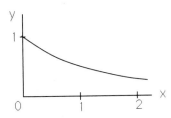

3.14. What is the resistance between the opposite corners A and B of the cube on the next page, whose edges are 1-ohm resistors?

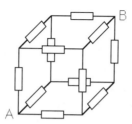

3.15. What is the resistance between the input terminals A and B of the infinite ladder network shown below, each of whose resistors is 1 ohm?

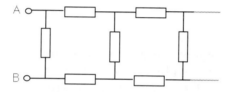

3.16. You have two straight iron bars that look identical. One is a magnet and one is not. How can you tell which is which by just touching them to each other?

3.17. The magic Fibonacci number appears in the network below. If each resistor is one ohm, what is the resistance $R$ between A and B?

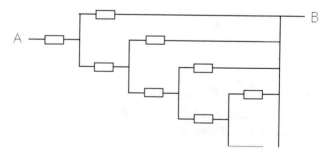

# Little Engineering Problems

3.18. Drop a ball from a height $h$ (for convenience, 4.905 m) onto the floor and let it bounce. The coefficient of restitution $e$ (the ratio of velocities of the ball just after and before hitting the floor) is 0.6. How long after it is dropped does the ball come to rest?

3.19. How far does the ball in problem 3.18 travel up and down before it comes to rest?

3.20. Weights with masses of 10 and 30 kg hang on the lightweight, well-oiled block and tackle shown below. Which way does the 30-kg mass move, and with what acceleration?

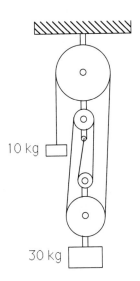

3.21. At Disney World a canal joining two lakes passes over a road on an aqueduct. When a boat goes across, does it increase the load on the aqueduct?

3.22. While a ship is floating in a closed canal lock, some of the cargo slides overboard and sinks to the bottom of the lock. Does the water level in the lock rise, fall, or remain the same?

3.23. A lightweight boat is floating in calm water. A man walks from one end of the boat to the other. Does the boat move while he is walking, and then stop; continue moving; or not move at all?

3.24. Shown below is a float attached by a string to the bottom of a jar filled with water. When the jar is accelerated, say to the left, does the float swing to the left in the jar, to the right, or stay put?

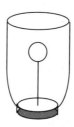

3.25. The three objects in the following figure are a glass ball, a glass ball with a small hole near the bottom, and a rubber balloon. All three are filled with air, and a weight is attached to each. They are lowered to the bottom of a pool where the weights are adjusted to give them neutral buoyancy. Then all three are raised halfway to the surface and released. What does each do: rise, sink, or remain where it is?

# Little Engineering Problems

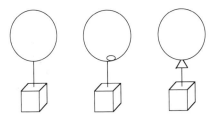

3.26. Here's a weird thought: Suppose that for some reason the earth's rotation started to speed up gradually. We would notice that the days were getting shorter and we were weighing less. How long would the day be when the people on the equator started to float up into the air? The radius of the earth at the equator is 6379 km and the force of gravity there at present is 9.78 N/kg.

3.27. Each resistance in the infinite network below is $r$. What is the resistance between A and B?

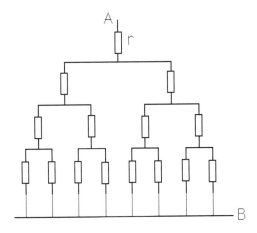

3.28. This one is about as far from reality as you can

get, but it's a good problem. A construction company is going to build a pipeline around a great circle of the earth. The design engineer mistakenly adds five meters to the length of the pipe. "That's OK," says the customer. "The pipe won't lie on the ground now, so build supports to hold it up at the same height all around the earth." Ignoring little things like mountains and oceans, calculate how high the supports must be.

3.29. The resistances in ohms in the figure below form three geometric progressions extending to infinity. What is the resistance between A and B?

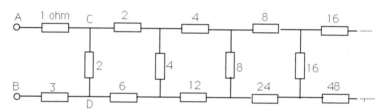

3.30. The network of resistances $R$ below has a *finite* number of sections. What is the characteristic resistance of the network, i.e., the resistance $R_0$ between the terminals C and D that makes the input resistance between A and B independent of the number of sections?

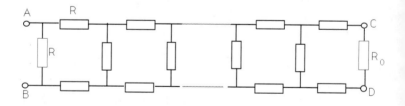

# Little Engineering Problems

3.31. The network of one-ohm resistors below extends to infinity in both directions. What is the resistance between points A and B?

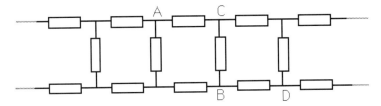

3.32. Lay a knife across a plate as in the figure below. If you press down with a force $F$ at the left end of the blade, the knife will tilt up but the plate will not move. If you shorten the distance $x_f$ and press down on the knife near the plate, the knife and plate will tilt together. At what value of $x_f$ does the crossover occur? Calculate the critical value of $x_f$ as a function of the weights of the knife and plate, $W_k$ and $W_p$, and the dimensions $a$, $b$, and $x_k$.

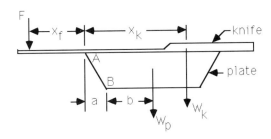

3.33. Shown next is a simple pin-connected truss supporting the weight $W$. What angle $\Theta$ does the arm A make with the horizontal if the tension in it, $F_A$, is twice the weight $W$? While you're at it, calculate the compressive force in arm B.

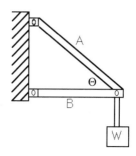

3.34. Below are two black boxes. Box A contains the Thevenin equivalent of some linear circuit, and Box B contains the Norton equivalent of the same circuit. Both boxes produce the same open-circuit voltage and the same short-circuit current. With access to only the outsides of the boxes and their terminals, how can you tell which is which?

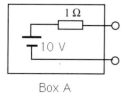
Box A

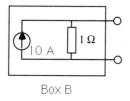
Box B

3.35. Two spheres of 10-cm diameter and 1-kg mass are out in space, motionless with respect to each other. Their centers are one meter apart. The only force on each sphere is the gravitational pull of the other sphere. How long will it be until they touch? Before you look at the answer, take a guess. You are clever if you are off by less than an order of magnitude.

3.36. In the infinite mesh of one-ohm resistors below, what is the resistance between any two adjacent nodes?

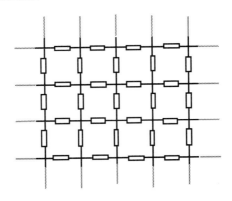

3.37. A candidate for the chair of her IEEE student branch read in the newsletter that a candidate must

1. be a student member of IEEE, not a senior, not on academic probation, and not active in branch functions, or

2. be a senior who is a student member of IEEE and not on probation, or

3. be a senior who is a student member of IEEE, not active in branch functions, and not on probation, or

4. be a non-senior who is a student member of IEEE, not on probation, and active in branch functions.

"Good grief," she said. "I'll draw a Karnaugh map and find out what that means." What was her conclusion?

3.38. How can you measure the internal resistance of a battery? Explain at least one way that won't discharge the battery quickly.

3.39. In the circuit below, the three resistances are equal. What are they?

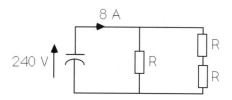

3.40. Below are the top and front views of an object. Please draw the side view.

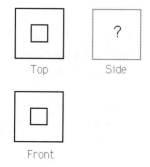

3.41. Here are all three views of another object, but the side view is incomplete. Please complete it.

**Little Engineering Problems** 53

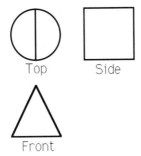

3.42. On the printed-circuit board below, connect terminals 1 and 1', 2 and 2', etc., without making any crossings.

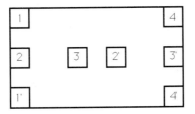

3.43. What is the famous signal at the output of the gate below?

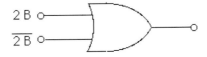

3.44. A sample of cork floats in water with 0.2 of its volume submerged and 0.8 out of the water. If you made a solid sphere of 2-meter diameter out of the cork, what would be needed to lift it: a small child, two football players, or a fork truck? You may add handles as needed.

3.45. This question is for engineers who have had a first course in linear systems. Can the frequency spectrum of the output of a time-invariant linear system contain a frequency not present in the spectrum of the input?

3.46. Suppose you are at the equator, standing on the top of a tower 100 meters high. Directly below you is a target on the ground. You hold a small weight directly over the bull's-eye and let it go. If there is no wind, will the weight land on the bull's-eye or is it certain to miss?

3.47. The two cylinders of radius $R$ and mass $m$ shown in the figure below are released at the same time and allowed to roll down a long 30-degree ramp. The lower cylinder is solid wood. The upper cylinder is an empty can made of thin steel. Will the steel cylinder keep up with the wooden cylinder, catch up to it, or fall farther behind? To verify your guess, calculate the acceleration of each cylinder.

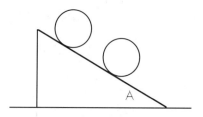

3.48. This problem will challenge your understanding of kinetics. A conveyor for unloading boxes from a truck consists of a series of steel rollers, as shown in the figure on page 56. The advantage of the rollers is that unless the ramp angle

# Little Engineering Problems

$A$ is too steep, the box moves down the ramp at a constant velocity. See if you can derive a formula for the velocity $v$ in terms of these parameters:

- $f$ = coefficient of sliding friction between a roller and the box
- $J$ = polar moment of inertia of a roller
- $L$ = spacing of the rollers' axes
- $M$ = mass of the box
- $m$ = number of rollers under the box that are not yet up to speed
- $n$ = number of rollers under the box
- $R$ = radius of a roller.

To simplify the problem, make these assumptions:

1. The weight of the box is evenly distributed among the rollers under it.

2. The velocity of the box does not fluctuate.

3. There is no friction in the bearings of the rollers.

4. When a roller's tangential velocity is up to the speed of the box, the rolling friction between them is negligible.

5. The number of rollers under the box is an integer.

6. Each roller is stationary when the box reaches it.

Then with the following numbers, please calculate the velocity $v$.

$A$ = 15 degrees
$f$ = 0.3
$J$ = 0.0125 kg·m²
$L$ = 12 cm
$M$ = 10 kg
$n$ = 6
$R$ = 4 cm

For extra credit, calculate the steepest ramp angle $A$ for which the boxes move down the ramp at a constant velocity.

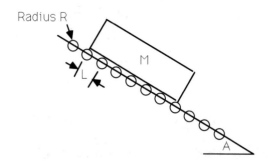

3.49. A cylindrical oil drum, whose diameter is 50 cm and whose height is 100 cm, is filled with oil and standing on the flat bed of a pickup truck. How much acceleration or deceleration of the truck can the drum withstand without tipping over? The bed of the truck has a rubber coating so the drum will not slide.

3.50. When the waves on the ocean approach the beach with the velocity $V$ at an angle $A$ in the

# Little Engineering Problems

figure below, the breaking crests move parallel to the beach with the velocity $V/\sin A$. Surfers like to ride those crests, because their velocity theoretically approaches infinity as $A$ approaches zero. How can this be? How could the crests move faster than the speed of light?

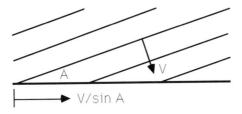

# Solutions to the Problems

## CHAPTER 1  PROBLEMS REQUIRING ONLY LOGIC

1.1. The bookworm eats through 36 mm of *Potentials*, as shown:

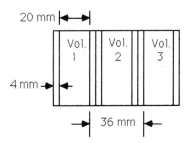

1.2. There are two ways to get four liters of water into the five-liter bottle.

- Fill the three-liter bottle and pour its contents into the five-liter bottle.

**Solutions to the Problems**

- Refill the three-liter bottle and pour its contents into the five-liter bottle to fill the latter, leaving one liter in the smaller bottle.

- Empty the five-liter bottle and pour into it the one liter from the smaller bottle.

- Refill the three-liter bottle and pour its contents into the larger bottle, bringing the level to four liters.

Or,

- Fill the five-liter bottle and use it to fill the three-liter bottle, leaving two liters in the larger bottle.

- Empty the smaller bottle and pour the two liters into it.

- Refill the five-liter bottle and use it to fill the smaller bottle, leaving four in the larger.

1.3. To identify the truth-teller, liar, and random answerer, first ask any of them, "Which of the other two lies more often?" Then ask the person indicated the same question. If the two people have pointed to each other, you know they are the truth teller and the liar. If not, you know that the first person you asked was the random answerer. To distinguish between the truth teller and the liar, just ask either of them the third question: "Is five equal to five?"

1.4. Number the sets of coins from 1 to 10. Remove

a number of coins from each set equal to the set's number, and weigh these coins all together. The number of grams less than 550 is the number of the set with nine-gram coins.

1.5. Since Bill's sock drawer contains two different colors of socks, he has to pick only three socks to make sure of getting a matched pair.

1.6. Below is the sequence of the knight's 63 moves. S and F mean start and finish, respectively.

```
 S 23 12 41 10 21 60  F
13 40  1 22 59 62  9 20
 2 37 24 11 42 51 56 61
39 14  3 52 55 58 19  8
36 25 38 15 50 43 54 57
31 28 33  4 53 46  7 18
26 35 30 49 16  5 44 47
29 32 27 34 45 48 17  6
```

1.7. Nobody said the four lines could not extend beyond the dots:

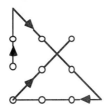

1.8. Nobody said the six sticks had to lie in the same plane. Construct a three-sided pyramid with an equilateral base:

1.9. The sum of what the engineers paid and the desk clerk pocketed doesn't mean anything. Of the $42 that the engineers paid, $40 was for the rent and $2 for the clerk.

1.10. Only one politician is honest, and 99 are crooked.

1.11. If George were type A, he wouldn't ask the question. If he is type B, the answer is "no," so Nancy must be type A.

1.12. One way to solve this problem is to consider all four possibilities for Anna: sane human, insane human, sane vampire, and insane vampire, and write down for each case what her answer requires Betsy to be. Then do the same for Betsy, and find the statements in the two lists that don't contradict each other. There are two pairs, both of which show that Anna is the vampire. The sisters are either both sane or both insane.

1.13. The coins are a half dollar and a nickel.

## Problems Requiring Only Logic

1.14. How did anyone in 44 BC know that the year was 44 BC?

1.15. Divide the nine coins into groups of three. Compare the weights of any two of the groups. If they balance, the counterfeit coin is in the third group. If not, the counterfeit coin is on the lighter side. Now you know which group contains the counterfeit coin. Compare the weights of any two coins in that group. If they balance, the third coin is the counterfeit. If not, the lighter coin is the counterfeit. A little inductive reasoning shows that you can identify one counterfeit among 27 coins with three weighings, one among 81 with four weighings, and one among $3^n$ coins with $n$ weighing.

1.16. The two clocks will come together again when they both show six o'clock. At ten seconds per hour, the drift will require 90 days. The time between dates in January and May can be 90 days only if the dates are January 31 and May 1, and it is not a leap year. Since it's the year of Harry's 23rd birthday and he was born in 1964 or 1965, the year is 1987 or 1988. But 1988 was a leap year, so Harry was born in 1964, and he is therefore the older.

1.17. Here are the seven permutations of red and white hats on the blind student (B) and the other two students (1 and 2):

|   | 1 | 2 | B |
|---|---|---|---|
| 1 | W | W | W |
| 2 | R | W | W |
| 3 | W | R | W |
| 4 | R | R | W |
| 5 | W | W | R |
| 6 | R | W | R |
| 7 | W | R | R |

If student 1 had seen two red hats, he would have known his hat was white. So the smart student B eliminated case 7. If student 2 had seen two red hats, she would have known hers was white, so case 6 was out. If she had seen a white hat on student 1 and a red hat on student B (WWR or WRR), she would have known hers was white because case 7 had been eliminated. So the blind student eliminated case 5. He realized that in each of the remaining four cases his hat was white. I think you'll agree the winner deserved to be the valedictorian.

1.18. If Mr. A were either a truth teller or a liar, his answer to Mr. B's question in their native language was, "I am a truth teller." So Mr. B is a liar.

1.19. Don't calculate the length of the bee's back-and-forth flight. Just observe that it flew for one hour, and therefore flew 60 km.

1.20. The oarsman's velocity relative to the water is constant. Since he rowed away from the log upstream for an hour, he rowed downstream

**Problems Requiring Only Logic**  65

for an hour before he returned to it. During those two hours the log moved two km, so its velocity relative to the land was 1 km/hr.

1.21. The big lumberjack is the little lumberjack's mother.

1.22. Instead of one monk climbing the hill one day and returning the next, consider two men, one going up and one coming down on the *same* day, both on the same path and walking at the same speed as the monk. Where they meet is the place that the monk passed at the same time going both ways.

1.23. There were $2n - 1$ responses to the host's question, all different. Since each person knew himself or herself and his or her spouse, the highest possible number of shakes by one person was $2n - 2$. Hence, the number of shakes had to be 0, 1, 2, ..., $2n - 2$. (That's $2n - 1$ responses.) The person who knew no one but his or her spouse and shook $2n - 2$ hands had to be the spouse of the person (A) who knew everyone and shook 0 hands, because everyone else saw at least one stranger: A's spouse. The hostess wasn't person A because someone else's spouse shook $2n - 2$ hands. The person who shook $2n - 3$ hands was the spouse of the person (B) who shook one hand because everyone else saw at least two strangers. Again, the hostess couldn't be person B because someone else's spouse shook $2n - 3$ hands. An extension of this reasoning produces this chart:

| The person shaking this number of hands | was the spouse of | the person shaking this number of hands. |
|---|---|---|
| 0 | | 2n - 2 |
| 1 | | 2n - 3 |
| 2 | | 2n - 4 |
| • | | • |
| • | | • |
| n - 2 | | n |

The remaining person is the hostess, who shook $n - 1$ hands, thereby completing the $2n - 1$ different numbers of handshakes. You can verify the solution by choosing a value for $n$ and drawing dots representing the people, in a circle for example. Draw lines between the dots to represent the handshakes. This procedure shows that the host also shook $n - 1$ hands.

1.24.

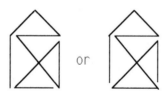

1.25. Ask, "Which side of the city do you live in, this side or the other side?" If you are on the east side, either a truth-telling easterner or a lying

**Problems Requiring Only Logic**

westerner will answer, "This side." If you are on the west side, either will answer, "The other side."

1.26. The question is, "Which road would a member of your family tell me is the road to Millinocket?" Either a truth teller or a liar will point to the correct road.

1.27.
1. LETTERS of the ALPHABET
2. WONDERS of the ANCIENT WORLD
3. ARABIAN NIGHTS
4. SIGNS of the ZODIAC
5. CARDS in a DECK (including the JOKERS)
6. PLANETS in the SOLAR SYSTEM
7. KEYS on a PIANO
8. STRIPES on the AMERICAN FLAG
9. DEGREES FAHRENHEIT at which WATER FREEZES
10. HOLES on a GOLF COURSE
11. DEGREES in a RIGHT ANGLE
12. DOLLARS for PASSING GO in MONOPOLY
13. SIDES on a STOP SIGN
14. BLIND MICE (SEE HOW THEY RUN)
15. QUARTS in a GALLON
16. HOURS in a DAY
17. WHEEL on a UNICYCLE
18. HEINZ VARIETIES
19. PLAYERS on a FOOTBALL TEAM
20. WORDS that a PICTURE is WORTH
21. DAYS in FEBRUARY in a LEAP YEAR
22. SQUARES on a CHESSBOARD
23. DAYS and NIGHTS of the GREAT FLOOD

1.28. You need a bag with one coin in it, a bag with two coins, but not a bag with three coins. You need a bag with four coins. (Two bags with two coins each instead of one bag with four coins won't work because you would run out of bags.) Before long you realize that the numbers of coins in the bags are powers of two: 1, 2, 4, 8, 16, 32, 64, 128, 256, and 512.

1.29. Did you see that this problem has the same solution as the preceding one? If the bills had binary numbers, their denominations would be 1, 10, 100, 1000, etc. IMUs. Then if a purchase price also were a binary number, you could tell just by the position of the 1's in that number which bills you would need.

1.30. The answer: name

1.31. The engineer could remove one lug nut from each of the other three wheels of his car and use them for the wheel with the flat tire.

1.32. Number the bottles 1 through 6. Then take one pill out of bottle 1, two out of bottle 2, etc., and weigh all 21 pills together. If the pile weighs $n$ grams, bottle number $n - 21$ is the one containing the two-gram pills.

1.33. The following table shows the moves that fulfill the conditions and get everyone across the river.

**Problems Requiring Only Logic** 69

| Trip no. | Row over | Left on near side | Row back | Left on far side |
|---|---|---|---|---|
| 1 | G,W | K,Q,P,PR | G | W |
| 2 | Q,PR | K,P,G | W | Q,PR |
| 3 | K,P | G,W | K,Q | P,PR |
| 4 | G,W | K,Q | P,PR | G,W |
| 5 | K,P | Q,PR | W | G,K,P |
| 6 | Q,PR | W | G | K,Q,P,PR |
| 7 | G,W | no one | | K,Q,P,PR,G,W |

1.34. The sequence is the first letters of the numbers One, Two, Three, etc., so the next two letters are E and N.

1.35. The symbols are the number 1 through 5 and their mirror images. Here is the sequence including the next two symbols:

1.36. The map on the next page shows that if the burro has to go down to the river, the trip to the mine would be the same length if the mine were three km *south* of the river. That trip would be the shortest along a straight line. A little geometry shows that the miner should water the burro at a point on the river 9.80 km east of his cabin.

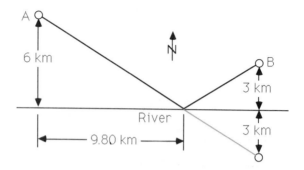

1.37. Remove the fourth and eleventh paper clips from the chain.

1.38. The fastidious fellow must have at least 15 shirts. Each Friday morning he puts one on, drops off seven, and picks up seven.

1.39. The student had three books: one with a blue cover, one with a red cover, and one with a green cover.

1.40. One sample from the can labeled "NUTS AND BOLTS" is enough. If it is a nut, that can is the true nuts can. The can labeled "BOLTS" can't contain bolts, so it must contain nuts and bolts. The remaining can, labeled "NUTS," contains bolts. If the sample is a bolt, the reasoning is similar.

1.41. The largest amount of change is $1.19: three quarters (or a half dollar and a quarter), four dimes, and four pennies.

1.42. This table shows that the senior is the oldest, shortest, and second lightest:

# Problems Requiring Only Logic

| Age | Height | Weight |
|-----|--------|--------|
| Sr  | Jr     | So     |
| Jr  | So     | Fr     |
| So  | Fr     | Sr     |
| Fr  | Sr     | Jr     |

1.43. In the impractical array of chairs below there are five rows of four chairs each.

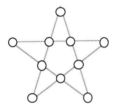

1.44. Here's how to cut a round pizza into eight pieces with three cuts:

1.45. Since sides A, B, D, and E are adjacent to side C, side F is opposite side C.

1.46. When reassembled, the four pieces of the square don't form a perfect rectangle. In the accurate drawing following there are two diagonal lines, each having two slopes. The slope of the segment forming the top of the lower trapezoid is 0.4. The slope of the hypotenuse of the lower triangle is 0.375. The thin empty space bounded by these two segments and the corresponding two above them is the extra one square unit.

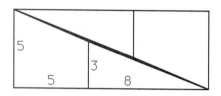

1.47. The cardboards were stacked in an orderly clockwise spiral:

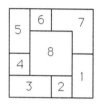

1.48. It's a small world after all

1.49. Unfinished symphony

1.50. Five feet underground

1.51. Just between you and me

1.52. I understand

1.53. Long underwear

1.54. Tricycle

1.55. Paradise

## Problems Requiring Only Logic

1.56. Line up in alphabetical order

1.57. Sailing, sailing over the seven seas

1.58. A bad spell of weather

1.59. Split level

1.60. The odds are overwhelming

1.61. He is an exponent of capitalism

1.62. See-through blouse

1.63. A little misunderstanding between friends

1.64. A backward glance

1.65. Three degrees below zero

1.66. Three degrees below zero

1.67. Once over lightly

1.68. Good afternoon

1.69. Different strokes

1.70. Age before beauty

1.71. Big man on campus

1.72. Getting it all together

1.73. Deer crossing

## Solutions to the Problems

1.74. Reading between the lines

1.75. Touchdown

1.76. Split-level house

1.77. The third child is John.

1.78. Here are the volumes of water in each container after each step:

| Beaker | Test tubes | | |
|---|---|---|---|
| 54 | 0 | 0 | 0 |
| 0 | 18 | 18 | 18 |
| 18 | 18 | 18 | 0 |
| 36 | 18 | 0 | 0 |
| 36 | 6 | 6 | 6 |
| 42 | 6 | 6 | 0 |
| 48 | 6 | 0 | 0 |
| 48 | 2 | 2 | 2 |
| 50 | 2 | 2 | 0 |

1.79. This table shows the volumes of solution in each of the three bottles after each pouring:

| 8 cc | 5 cc | 3 cc |
|---|---|---|
| 8 | 0 | 0 |
| 3 | 5 | 0 |
| 3 | 2 | 3 |
| 6 | 2 | 0 |
| 6 | 0 | 2 |
| 1 | 5 | 2 |
| 1 | 4 | 3 |
| 4 | 4 | 0 |

**Problems Requiring Only Logic** 75

1.80. The chart below shows that all of the glasses can be inverted in six moves. The glasses inverted in each move are shaded.

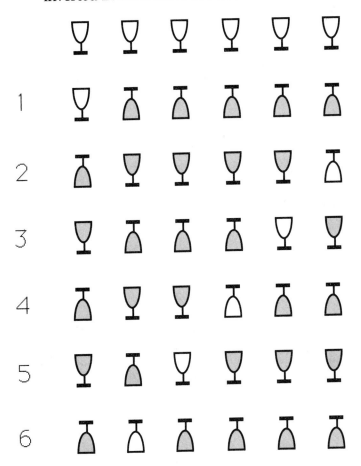

1.81. The figures on the next page show three steps of a solution that requires 22 strokes.

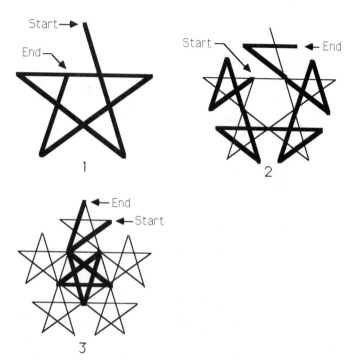

1.82. The obvious answer is that it doesn't matter which of the two closed doors you choose. The probability that the Miata is behind either of them is 1/2, right? You can get this result by intuition or by applying Bayes' theorem. Unfortunately, it's wrong. Here's the correct solution: If your strategy is *not* to switch, it doesn't matter whether the host opens a door or not, and your probability of winning the car is 1/3. If your strategy is to switch, you'll lose only if you originally choose the door concealing the Miata. If you choose either of the other two doors, you'll win. So your probability of winning is 2/3. So the answer is: switch your choice!

# Problems Requiring Only Logic

1.83. Without switching, you win if you choose the door concealing the Miata, and lose if you choose another door. So if there are $n$ doors, your probability of winning is $1/n$. With switching, you lose if you originally choose the door concealing the Miata, and win if you originally choose any of the other doors. So your probability of winning is $(n-1)/n$.

1.84. The figure below shows that 11 open triangles can be formed with seven straight lines.

1.85. If you can understand the question, you see that the answer is "yes."

1.86. Make two columns of the integers from 1 to 100, one counting up from 1 and the other counting down from 100, like this:

$$\begin{aligned} 1 + 100 &= 101 \\ 2 + 99 &= 101 \\ \cdots \\ 99 + 2 &= 101 \\ 100 + 1 &= 101 \end{aligned}$$

Then add the numbers in the two columns horizontally. The third column shows that the sum of the integers from 1 to 100 is

$$\frac{(100)(101)}{2} = 5050 .$$

In general, the sum of the integers from 1 to $n$ is

$$\frac{n(n+1)}{2} .$$

With this formula you could make the calculation before the teacher could get up from his chair.

1.87.

| Size of square | Squares in row | Squares in column | Total |
|---|---|---|---|
| 1 x 1 | 8 | 8 | 64 |
| 2 x 2 | 7 | 7 | 49 |
| 3 x 3 | 6 | 6 | 36 |
| 4 x 4 | 5 | 5 | 25 |
| 5 x 5 | 4 | 4 | 16 |
| 6 x 6 | 3 | 3 | 9 |
| 7 x 7 | 2 | 2 | 4 |
| 8 x 8 | 1 | 1 | 1 |
| | | Total | 204 |

# CHAPTER 2 PROBLEMS REQUIRING SOME MATHEMATICS

2.1. Assume that
$$e^\pi > \pi^e .$$
Then

$$\pi > e \ln \pi$$
and
$$\frac{\pi}{\ln \pi} > e = \frac{e}{\ln e}.$$

For any variable $x$,

$$\frac{d}{dx}\left[\frac{x}{\ln x}\right] = \frac{\ln x - 1}{(\ln x)^2},$$

which is positive for any $x$ greater than $e$. Since $\pi$ is greater than $e$, Inequality (1) is true, and the assumption was correct. In fact, $e^\pi = 23.14$ and $\pi^e = 22.46$.

2.2. If the engineer runs to the north end of the tunnel, he travels the distance $L/4$ while the car travels the distance $x$. If he runs to the south end, he travels $3L/4$ while the car goes $x + L$. Subtracting shows that he could run the distance $L/2$ while the car goes $L$. His speed is therefore 20 mph, or three minutes per mile, pretty good time if he is wearing hiking boots.

2.3. The missing number is 31. The numbers in the series are the decimal base number 16 in all the number bases starting with base 16 and ending with base 2.

2.4. If the distance from Los Angeles to San Francisco is $D$, the total time for the round trip was

$$\frac{D}{380} + \frac{D}{420} = \frac{D}{199.5}.$$

The average speed for the round trip was therefore

$$\frac{2D}{D/199.5} = 399 \text{ mph},$$

not 400 mph.

2.5. Square both numbers. Notice that

$$(\sqrt{17} + \sqrt{10})^2 = 27 + 2\sqrt{170}$$

and

$$(\sqrt{53})^2 = 27 + 26 = 27 + 2\sqrt{169},$$

so $\sqrt{17} + \sqrt{10}$ is larger than $\sqrt{53}$.

2.6. The smallest number identified so far is

105,263,157,894,736,842.

Half of it is

52,631,578,947,368,421.

2.7. Assume that the births of daughters are independent events, each having a probability of 0.5. The probability of having five consecutive daughters is then

## Problems Requiring Some Mathematics

$$(0.5)^5 = 0.031 .$$

2.8. The reliability of the guidance system is

$$(0.999)^{100} = 0.905 .$$

2.9. In an honest crap game the thrower has a probability of winning of 0.493. So bet against the guy with the dice.

2.10. The probability that two people have different birthdays is 364/365. The probability that three people have different birthdays is

$$P = \left[\frac{364}{365}\right]\left[\frac{363}{365}\right] .$$

The probability that $n$ people have different birthdays is

$$P = \left[\frac{364}{365}\right]\left[\frac{363}{365}\right] \cdots \left[\frac{366-n}{365}\right] .$$

When $n$ is 23, $P = 0.493$. So for a group of 23 people the probability 1-$P$ that at least two have the same birthday is slightly higher that 0.5.

2.11. If the traffic light is red for 30 seconds and green for 30 seconds, and the random variable **x** is the duration of waiting, its density function $f(x)$ is as shown below. The arrowhead indicates an impulse with an area of 0.5. The expected wait is

$$E(\mathbf{x}) = \int_{-\infty}^{\infty} x f(x)\, dx = \frac{1}{60} \int_0^{30} x\, dx = 7.5 \text{ seconds}.$$

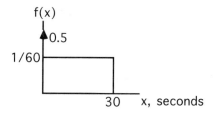

2.12. If you go first, your probability of winning on your $k+1$th throw is

$$(1-p)^k (1-p)^k\, p = (1-p)^{2k}\, p\,.$$

Your probability of winning on *some* throw is

$$P(A) = p \sum_{k=0}^{\infty} (1-p)^{2k}\,.$$

Your friend's probability of winning on his $k+1$th throw is

$$(1-p)^k (1-p)^{k+1}\, p = p\,(1-p)(1-p)^{2k},$$

so his chance of winning the game is

$$P(B) = p\,(1-p) \sum_{k=0}^{\infty} (1-p)^{2k} = (1-p)\, P(A)\,.$$

Since

$$P(A) + P(B) = 1,$$

$$P(A) = \frac{1}{2-p} \quad \text{and} \quad P(B) = \frac{1-p}{2-p}.$$

It's important to take the first turn. If you're flipping a coin, the one who goes first has twice as good a chance of winning.

2.13. What is the probability that the three pieces formed by cutting a straight stick at random in two places will form a triangle? An equivalent statement is, "What is the probability that none of the three pieces is longer than half the length of the stick?" Let the length of the stick be 1. Let $x$ and $y$ be the distances from one end of the stick to the two cuts. Since $x$ and $y$ are each equally likely to be anywhere between 0 and 1, the joint probability density of $x$ and $y$ is uniform, and the point $x,y$ in the next figure is equally likely to lie anywhere within the large square. If the three pieces can form a triangle, $x$ and $y$ cannot both be less than 0.5, nor can both be greater than 0.5, and their difference $x$-$y$ or $y$-$x$, whichever is positive, must be less than 0.5. That leaves only the hatched area in the figure. The probability that the pieces can form a triangle is the ratio of the hatched area to the total area, or 0.25.

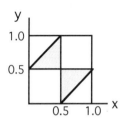

2.14. All of the sums of three integers whose product is 72 are listed here:

|   |   |    | Sum |
|---|---|----|-----|
| 1 | 1 | 72 | 74  |
| 1 | 2 | 36 | 39  |
| 1 | 3 | 24 | 28  |
| 1 | 4 | 18 | 23  |
| 1 | 6 | 12 | 19  |
| 1 | 8 | 9  | 18  |
| 2 | 2 | 18 | 22  |
| 2 | 3 | 12 | 17  |
| 2 | 4 | 9  | 15  |
| 2 | 6 | 6  | 14  |
| 3 | 3 | 8  | 14  |
| 3 | 4 | 6  | 13  |

Since the guest asked for more information, the house number must have been 14. As soon as he heard "The oldest girl ...," he knew their ages were 8, 3, and 3.

2.15. In order to differentiate the equation

$$x^2 = x \cdot x = x + x + x + \ldots$$

**Problems Requiring Some Mathematics** 85

we have to specify the number of $x$'s on the right side. That makes one of the $x$'s in $x \cdot x$ a constant, so the derivative of the right side is

$$1 + 1 + 1 + \ldots (x \text{ times}) = x,$$

not $2x$.

2.16. 1 is the smallest integer divisible by 1; 2 is the smallest integer divisible by 1 and 2; 6 is the smallest integer divisible by 1, 2, and 3; etc. The next number in the sequence, following 840, is 2520, the smallest integer divisible by 1, 2, 3, ..., 9.

2.17. Take the pennies! For the last day alone you'll earn $2^{30}$ cents or \$268,435,456.

2.18. The numbers in the squares below add up to 15 horizontally, vertically, and diagonally.

| 4 | 9 | 2 |
|---|---|---|
| 3 | 5 | 7 |
| 8 | 1 | 6 |

2.19. To swap A and B, write

$$A = A - B$$
$$B = A + B$$
$$A = B - A$$

2.20. Solving this problem is easy if you know how to use Bayes' theorem. The engineer is right: the probability that the ball left in the box is black is 2/3.

2.21. Let $g(n)$ be the payoff for $n$ consecutive tails, and $p(n)$ be the probability of flipping them. The expected payoff of the game is

$$E[g(n)] = g(1)p(1) + g(2)p(2) + g(3)p(3) + ...$$
$$= (1)(1/2) + (2)(1/4) + (3)(1/8) + ...$$
$$= \sum_{n=1}^{\infty} n/2^n = 2 .$$

The fair flipper's fee is therefore two dollars.

2.22. Now the expected payoff is

$$E[g(n)] = (2)(1/2) + (4)(1/4) + (8)(1/8) + ...$$
$$= 1 + 1 + 1 + ... = \infty .$$

It doesn't make sense intuitively that the fair entrance fee is infinite, but the calculation doesn't lie. The house would never establish this game.

2.23. As shown below, the rope touches 120 degrees or $\pi/3$ meters of each barrel, so the length of the rope is

$$(3)(1 + \pi/3) = 3 + \pi \text{ meters} .$$

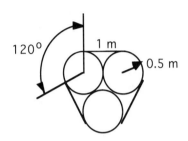

2.24. If the radii of the hole and the sphere shown below are $r_1$ and $r_2$, the hatched differential element of volume is

$$dv = (2\pi x)(2y)dx ,$$

and the volume of solid material is

$$v = \int_{x=r_1}^{x=r_2} dv = 4\pi \int_{r_1}^{r_2} x(r_2^2 - x^2)^{1/2} dx$$

$$= \frac{4}{3} \pi (r_2^2 - r_1^2)^{3/2} .$$

Since
$$r_2^2 - r_1^2 = 3^2 ,$$

$$v = 36\pi \text{ cm}^2 .$$

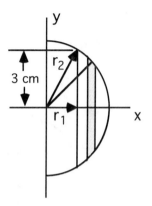

2.25.
1. If there is no wind, the round trip takes 2 hours.

2. If there is a 40 km/hr headwind, the trip out takes 200/160 or 5/4 hr. The return trip takes 200/240 or 5/6 hr. The round trip takes 25/12 or 2.08 hr.

3. If there is a 40 km/hr crosswind, the airplane's groundspeed is

$$(200^2 - 40^2)^{1/2} = 196 \text{ km/hr},$$

so the time for the round trip is 400/196 or 2.04 hr.

2.26. The probability that the coupon in the second box is different from the first is 4/5. So on the average a customer must buy 1 + 5/4 boxes to get two different

**Problems Requiring Some Mathematics** 89

coupons. Then he must buy 5/3 more boxes on the average to get three different coupons. The average number of boxes required to produce five different coupons is

$$1 + 5/4 + 5/3 + 5/2 + 5/1 = 11.4 \ .$$

Let's hope the customer likes the cereal.

2.27. Since each car is constantly aimed at the car ahead, and is moving perpendicular to it, they will collide after traveling one kilometer. Their speeds are unimportant as long as they are all equal.

2.28. Mary wins. If the distance around one lap is $d$ and Mary's speed is $v$, her time is $2d/v$. Bill's time is

$$\frac{d}{v-x} + \frac{d}{v+x} = \frac{2d}{v - \frac{x^2}{v}} \ ,$$

which is slower.

2.29. Place a sphere of unit radius at the intersection of the axes. Then make a section parallel to the plane of the axes, as shown in the next figure. The area in the section common to both cylinders is a square, and a circular section of the sphere is inscribed in it. The ratio of the areas of the square and the circle is $4/\pi$. Make a stack of such sections, each with a differential thickness. Their total volume is the volume common to the two cylinders, and the total volume of the circles of differential thickness is the volume of the sphere, $4\pi/3$. Since the ratio of the total volumes is the same as the ratio of the areas in each section, the common volume is $(4/\pi)(4\pi/3)$ or $16/3$.

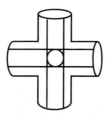

2.30. If $v_1$, $v_2$, and $v_3$ are the velocities of the first man, second man, and the ship, $L$ is the length of the ship, and $d$ is the distance between the two men when the ship has just passed the first man,

$$10(v_1 + v_3) = L,$$

$$9(v_2 + v_3) = L,$$

and

$$v_3 = \frac{d - L - 600v_2}{600}.$$

When the ship has just passed the first man, the time until they meet is

$d/(v_2 - v_1)$. Manipulation of the equations shows that

$$d/(v_2 - v_1) = 6090 \text{ seconds}.$$

Ten minutes and nine seconds later, the time remaining until the two men meet is 6090 - 609 or 5481 seconds.

2.31. The next row is 1113213211. Each row describes the one above it. The first row consists of 1 one, making the second row "11." The second row consists of two ones, making the third row "21." Since the third row consists of 1 two and 1 one, the fourth row is "1211," and so on.

A little thought shows that no number can appear more than three consecutive times in one row. Numbers 4 and higher will therefore not appear in the next row.

2.32. 0! + 0! + 0! = 1 + 1 + 1 = 3 .

2.33. Let $r_1$ and $r_2$ be the radii of the inner and outer circles, and $2r_3$ be the length of the chord, as in the next figure. Notice that

$$r_3^2 = r_2^2 - r_1^2.$$

The annular area is

$$A = \pi(r_2^2 - r_1^2) = \pi r_3^2 ,$$

where $r_3$ is half the length of the chord.

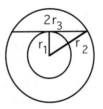

2.34. Whoever served first served five games. If he won $x$ of these, he lost $5 - x$. If the other player lost $y$ games while serving, the total lost by the servers is

$$5 - x + y = 5 .$$

Thus $x = y$, and the first server won $x + y$ or $2x$ games, an even number. Since the set score was 6 to 3, the first server was James.

2.35. For the parabolic section of the glass, shown in the next figure,

$$y = k r^2 .$$

The original volume of the dessert was

$$v_h = \int_0^h \pi r^2 \, dy = \frac{\pi}{k}\int_0^h y \, dy = \frac{\pi y^2}{2k}\bigg|_0^h = \frac{\pi h^2}{2k} .$$

The volume that Dorothea left for Michael was

$$v_x = \frac{\pi x^2}{2k}.$$

So if $v_x = v_h/2$,

$$x^2 = \frac{h^2}{2} \quad \text{or} \quad x = 0.707\ h.$$

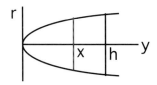

**2.36.** Notice that

$$y = \sqrt{\sin x + y}.$$

Just differentiate to show that

$$\frac{dy}{dx} = \frac{\cos x}{2y - 1}.$$

**2.37.** Notice that
$$z = \sqrt{2 + z}.$$

Thus

$$z^2 = 2 + z$$

and

**94**  **Solutions to the Problems**

$$z = 2 \text{ or } -1 .$$

2.38. Since $n - n = 0$, the value of the polynomial is zero.

2.39. Since $j = e^{j\pi/2}$,

$$j^j = e^{(j\pi/2)j} = e^{-\pi/2} = 0.208 .$$

Of course $j$ has many other values, such as $e^{j5\pi/2}$ and $e^{j3\pi/2}$, and therefore $j^j$ has many other values.

2.40. A. $(5+5+5+5)(5)$ or $(5)(5)(5)-(5)(5)=100$

B. The smallest integer is 1; for example, $2/2$.

C. $(123456789)^0 = 1$

D. $11^{11} = 2.853 \times 10^{11}$

2.41. If the $k$th term of the series is $x(k)$ and the term after it is $x(k+1)$, etc.,

$$x(k + 2) = x(k + 1) + x(k) .$$

The solution to this second-order, linear difference equation with constant coefficients is

$$x(k) = c_1 \left(\frac{1 + \sqrt{5}}{2}\right)^k + c_2 \left(\frac{1 - \sqrt{5}}{2}\right)^k,$$

where $c_1$ and $c_2$ are constants determined by the initial two terms in the series. It's easy to show that regardless of what the first two terms are,

$$\lim_{k \to \infty} \frac{x(k+1)}{x(k)} = \frac{1 + \sqrt{5}}{2} = 1.618...$$

2.42. The product of a column vector **a** and a row vector **b** is the matrix

$$\begin{bmatrix} a_1b_1 & a_1b_2 & \cdots & a_1b_n \\ a_2b_1 & a_2b_2 & \cdots & \cdot \\ \cdot & & & \cdot \\ \cdot & & & \cdot \\ a_nb_1 & \cdots & \cdots & a_nb_n \end{bmatrix}$$

which is singular and whose determinant is therefore zero.

2.43. The figure on the next page shows the original and modified pool bottoms. In the original, $a = 6$. Since the blue areas (hatched) of the two designs are equal, and the white areas are equal,

$$a^2 = 2b + 2c + 4$$

and

$$4a + 4 = bc.$$

The solution to these equations is

$$b = 14 \text{ or } 2,$$

and

$$c = 2 \text{ or } 14,$$

so the bottom of the revised pool will be 4 x 16 squares.

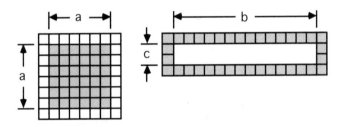

2.44. Wrong! Bill can get to class in $t$ hours at 15 km/hr or $t+2$ hours at 10 km/hr, so

$$15t = 10(t + 2),$$

and $t = 4$ hr, so the distance to the classroom is 60 km. To get there in five hours, he must average 12 km/hr, not 12.5.

2.46. If team A wins the first game, it must win three of the next six games to win the series. The number of ways to do that is the number of combinations of six things taken three at a time, or

$$\binom{6}{3} = 20.$$

The opponents, team B, must win four of the next six games in order to win the series. The number of ways to do that is

$$\binom{6}{4} = 15.$$

Of the 35 outcomes, 20 produce a win for team A, so its probability of winning the series is

$$P = \frac{20}{35} = \frac{4}{7}.$$

If team A wins the second game, its probability of winning the series is

$$P = \frac{\binom{5}{2}}{\binom{5}{2} + \binom{5}{4}} = \frac{10}{10 + 5} = \frac{2}{3},$$

and if it wins the third game,

$$P = \frac{\binom{4}{1}}{\binom{4}{1} + \binom{4}{4}} = \frac{4}{4 + 1} = \frac{4}{5}.$$

2.47. When you divide number A by number B, if A is one less than a multiple of B, the remainder is one less than B. For example, when 48 is divided by 7, the remainder is 6. So let's choose a number that is one less than a multiple of 2 through 9. A logical choice is $(2)(3)(4)\cdots(9) - 1$ or 362,879.

2.48. The solution is straightforward if you are familiar with Bayes' theorem. If $A_1$, $A_2$, and $A_3$ are the choices of boxes 1, 2, and 3, and $B$ is the draw of a black ball, the new probability that you have chosen box 1 is

$$P(A_1|B) = \frac{P(A_1)P(B|A_1)}{P(A_1)P(B|A_1)+P(A_2)P(B|A_2)+P(A_3)P(B|A_3)}$$

$$= \frac{\frac{1}{3}(1)}{\frac{1}{3}(1) + \frac{1}{3}(0) + \left(\frac{1}{3}\right)\left(\frac{1}{2}\right)} = \frac{2}{3}.$$

The additional information changed the probability!

2.49. If $A_1$ and $A_2$ are the events that bridge $A$ is down or up, and $B_1$ and $B_2$ mean the same for bridge $B$, etc., the probability that a car can cross the river is

$$P(C) = P(A_1D_1 + B_1E_1 + A_1C_1E_1 + B_1C_1D_1),$$

where, for example, $A_1D_1 + B_1E_1$ means "$A_1$ and $D_1$, or $B_1$ and $E_1$." The probability that a *boat* can pass through the bridges is

$$P(B) = P(A_2B_2 + D_2E_2 + A_2C_2E_2 + B_2C_2D_2).$$

If a car can cross the river, a boat can't pass, and vice-versa. Therefore $P(B) + P(C) = 1$. Since all of the probabilities $P(A_1)$, $P(A_2)$, $P(B_1)$, etc. are equal and independent,

$$P(C) = P(B) = 1 - P(C)$$

and

$$P(C) = 0.5 \ .$$

# CHAPTER 3  LITTLE ENGINEERING PROBLEMS

3.1. When the box is at rest in the water, the buoyant force of the water equals the force of gravity. When the box sinks a distance $y$ from its rest position, the additional upward buoyant force is $\rho Agy$, where $\rho$ is the density of the water, $A$ is the area of the box top or bottom, and $g$ is the force of gravity, 9.81 N/kg. The downward inertial force is $m\ddot{y}$, where $m$ is the mass of the box. Thus

$$m\ddot{y} + \rho A g y = 0 ,$$

or

$$y = k_1 \sin \sqrt{\frac{\rho A g}{m}} t + k_2 \cos \sqrt{\frac{\rho A g}{m}} t .$$

The period of oscillation is therefore

$$T = 2\pi \sqrt{\frac{m}{\rho A g}}$$

and the weight of the box is

$$m = \frac{\rho A g T^2}{4\pi^2} = \frac{(1025)(1)(9.81)(2)^2}{(4)(\pi)^2}$$

$$= 1019 \text{ kg} .$$

3.2. What is the gravitational force on a car in the tunnel when it is at the distance $y$ from the center of the earth? Consider the earth to be two concentric spheres. The inner is a solid sphere of radius $y$ and mass $m_y$, and the outer is a hollow sphere of inner radius $y$ and outer radius $R$. The outer sphere exerts no gravitational force on the car because the car is inside it. The inner sphere exerts the force

$$f = \frac{Gm_y m_c}{y^2}$$

on the car, where $G$ is the gravitational constant and $m_c$ is the mass of the car. If the density of the earth is assumed to be uniform,

$$m_y = \left[\frac{y}{R}\right]^3 m_e,$$

where $m_e$ is the mass of the earth. Therefore

$$f = \frac{Gm_e m_c}{R^3} y. \tag{1}$$

The inertial force on the car, $m_c \ddot{y}$, is equal and opposite to the gravitational force. Thus

$$\ddot{y} + \frac{Gm_e}{R^3} y = 0.$$

The solution to this differential equation is

$$y = c_1 \sin \sqrt{k}\, t + c_2 \cos \sqrt{k}\, t, \tag{2}$$

where

$$k = \frac{Gm_e}{R^3}.$$

At $t = 0$, $y = R$ and $dy/dt = 0$. Eq.(2) becomes

$$y = R \cos \sqrt{k}\, t.$$

When the car reaches the center of the earth where $y = 0$,

$$\cos \sqrt{k}\, t = 0, \quad \text{and} \quad \sqrt{k}\, t = \frac{\pi}{2}.$$

The time required is

$$t = \frac{\pi}{2\sqrt{k}} = \frac{\pi}{2} \sqrt{\frac{R^3}{Gm_e}}.$$

The time needed for the complete trip is twice this amount. To simplify the expression, observe that according to Eq.(1) the force on the car when it is at the surface is

$$f = m_c g = \frac{Gm_e m_c}{R^2},$$

where $g$ is the force of gravity there, 9.81 N/kg. Thus

$$\frac{R^3}{Gm_e} = \frac{R}{g},$$

and the time required for a passage through the tunnel is

$$2t = \pi\sqrt{\frac{R}{g}} = \pi\sqrt{\frac{6.371 \times 10^6}{9.81}}$$

= 2532 seconds or 42.2 minutes.

3.3. The centrifugal force on a low-level satellite of mass $m$ is $mR\omega^2$, where $R$ is the radius of the earth and $\omega$ is the angular velocity of the satellite around the earth. The gravitational force on the satellite is $mg$. Equating these two forces shows that the period of rotation of the satellite around the earth is

$$2\pi\sqrt{\frac{R}{g}} = 84.4 \text{ minutes,}$$

which is exactly twice the time required by the car in Problem 3.2 to fall through the earth. So a trip halfway around the earth, either around by satellite or through the hole, takes the same time, 42.2 minutes.

3.4. The centrifugal force on a geosynchronous satellite of mass $m$ at a distance $r$ from the center of the earth is $mr\omega^2$, where $\omega$ is the angular velocity of the satellite around the earth, one revolution per day. If the earth's radius is $R$, the gravitational force on the satellite is

$mgR^2/r^2$. Equating these two forces reveals that the satellite's altitude $r - R$ is 35,900 km.

3.5. If a meteorite is at the distance $y$ from the center of the earth and approaching with the velocity $v$, its acceleration toward the earth is

$$\frac{dv}{dt} = -\frac{gR^2}{y^2},$$

where $R$ is the radius of the earth. Now

$$\frac{dv}{dy} = \frac{dv}{dt}\frac{dt}{dy} = \frac{dv/dt}{dy/dt} = \frac{dv/dt}{v}$$

or

$$v\,dv = \frac{dv}{dt}\,dy = -gR^2\,\frac{dy}{y^2}.$$

Integrating both sides of this equation gives

$$\frac{v^2}{2} = \frac{gR^2}{y} + c.$$

Since $v = 0$ when $y = \infty$, $c = 0$ and

$$v = \sqrt{\frac{2gR^2}{y}}.$$

When the meteorite reaches the earth's

atmosphere at an altitude $y - R$ of 200 km, its velocity is 11.0 km/s.

3.6. If the height of a cylindrical can is $h$ and the diameter of its base is $d$, its volume is

$$V = \frac{\pi d^2 h}{4}$$

and its surface area is

$$A = \pi dh + \frac{\pi d^2}{2} . \qquad (1)$$

Hold the volume constant, solve for $d$ in terms of $V$, and substitute the result into Eq.(1) to eliminate $d$. Then differentiate $A$ with respect to $h$ to find the value of $h$ that minimizes $A$. Alternatively, you can eliminate $h$ in Eq.(1) and differentiate the latter with respect to $d$. Or, you can hold $A$ constant and maximize $V$. All of the methods yield the same result: $h = d$. One reason why cans are made with the ratio $h/d$ larger than 1 is to make them look larger.

3.7. If the elevation angle of a gun is $\Theta$, the vertical component of the bullet's velocity $v$ is initially $v \sin\Theta$. If air resistance is neglected, the time that is required for the bullet to reach its maximum height and fall back to the ground is $2v \sin\Theta/g$.

Since the horizontal component of the bullet's velocity is $v\cos\Theta$, it travels the horizontal distance $2v^2\sin\Theta\cos\Theta/g$. Differentiating this distance with respect to $\Theta$ shows that the elevation angle that maximizes the horizontal distance is 45 degrees.

3.8. The series is the list of standard five-percent resistor values.

3.9. The time required for the water to fall to the ground is

$$t = \sqrt{\frac{2(d-h)}{g}}.$$

The product

$$vt = 2c\sqrt{dh-h^2} \tag{1}$$

is the horizontal distance the water travels. Maximizing this distance with respect to $h$ shows that the water squirts farthest when $h = d/2$. Putting this value and $c = 1$ into Eq.(1) shows that the maximum squirting distance is $d$, the depth of water in the tank.

3.10. To account for the lost energy, redraw the circuit with a resistor $R$ included to represent the resistance of the wire, as shown below. If you derive and solve the differential equation for the current $i$

flowing after the switch is thrown to the right, you find that

$$i = \frac{V}{R} e^{-\frac{2}{RC} t}.$$

The energy dissipated by the resistor is

$$\int_0^\infty i^2 R \, dt = \frac{CV^2}{4},$$

the missing half of the original energy. Notice that it is independent of the resistance of the wire, if the resistance is any amount greater than zero.

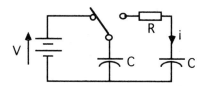

3.11. Draw a free-body diagram of each mass and write the sum of the forces on each mass. The sums are

$$m_1(g + a) = T$$

$$m_2(g - a) = T,$$

where $a$ is the acceleration of the masses and $T$ is the tension in the rope. Solving these equations shows that $a = g/2 = 4.90 \text{ m/s}^2$.

3.12. If the mass has an angular velocity ω fast enough to keep the string taut when it reaches its highest point after hitting the pin, the centrifugal force on the mass is at least equal to its weight:

$$mr\omega^2 = mg$$

or

$$\omega^2 = \frac{g}{r}.$$

The kinetic energy of the mass there is

$$\frac{mv^2}{2} = \frac{mr^2\omega^2}{2} = \frac{mgr}{2}.$$

The original potential energy of the mass $mgL$ has been converted to the kinetic energy $mgr/2$ and the potential energy $2mgr$. Thus

$$mgL = \frac{mgr}{2} + 2mgr = \frac{5mgr}{2},$$

and

$$r = \frac{2L}{5}.$$

The distance $d$ must therefore be at least $3L/5$.

## Little Engineering Problems

3.13. Yes, the cornucopia has an infinite inner surface area and encloses a finite volume. To prove these facts that seem contradictory, let $ds$ be a differential distance along the curve. The area of the cornucopia's surface is

$$A = \int_0^\infty 2\pi y \, ds > \int_0^\infty 2\pi y \, dx = \infty.$$

The volume enclosed by the cornucopia is

$$V = \int_0^\infty \pi y^2 dx = \pi \int_0^\infty \frac{dx}{(1+x)^2},$$

which is difficult to integrate. But notice that

$$V < \pi \int_0^1 \frac{dx}{(1+x)^2} + \pi \int_0^\infty \frac{dx}{x^2},$$

and both of the right-hand terms are finite, so the volume $V$ is finite.

3.14. The resistance between the opposite corners of the cube is 5/6 ohm.

3.15. Let the resistance between terminals A and B of the network, redrawn below, be $R$. The resistance to the right of terminals C and D is also $R$. So the former resistance is 1 ohm in parallel with $R+2$ ohms, or

$$R = \frac{R+2}{R+3},$$

and $R = 0.732$ ohms.

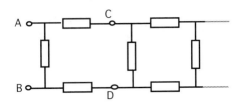

3.16. To tell which iron bar is a magnet, make a T with them. They stick together when the unmagnetized bar is the crossbar, but not when the magnet is the crossbar.

3.17. If the switch at C is open in the network, redrawn here, the resistance between C and B is $R$, the same as the resistance between A and B when the switch is closed. So

$$R = 1 + \frac{R}{1+R},$$

and $R = 1.618$ ohms, the magic Fibonacci number.

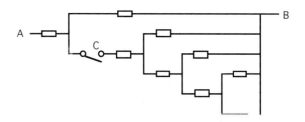

3.18. After falling the distance $h$, the ball hits the floor with the velocity

$$v = \sqrt{2gh}.$$

The time of the fall is

$$t = v/g = \sqrt{2h/g}.$$

The ball bounces up with an initial velocity $ev$. The duration of the first bounce, up and down, is therefore

$$2e\sqrt{2h/g}.$$

The time required for the second bounce is

$$2e^2\sqrt{2h/g}.$$

The time required for all of the bounces is

$$T = \sqrt{2h/g}\,(1 + 2e + 2e^2 + 2e^3 + \ldots)$$

$$= \sqrt{2h/g}\left[1 + \frac{2e}{1-e}\right] = 4 \text{ seconds}.$$

3.19. Since the ball starts the first bounce with the velocity $ev$, it reaches a height

$$(ev)^2/2g = e^2 h.$$

During the second bounce it reaches the height

$$(e^2 v)^2/2g = e^4 h.$$

After all of the bounces, the ball has traveled the total distance

$$h + 2h(e^2 + e^4 + e^6 + \ldots) = h\left[1 + \frac{2e^2}{1 - e^2}\right]$$

$$= 10.42 \text{ meters}.$$

3.20. The following free-body diagrams show the forces on the two masses, $m_1$ (10 kg) and $m_2$ (30 kg). The tension in the rope is $T$, and the assumed directions of the accelerations $a_1$ and $a_2$ are indicated. Since the tackle gives $m_2$ a mechanical advantage of four over $m_1$,

$$a_1 = 4a_2,$$

and the sums of forces on the two masses are

$$m_1(g - a_1) = m_1(g - 4a_2) = T$$

and

$$m_2(g + a_2) = 4T.$$

Thus the acceleration of the 30-kg mass is

$$a_2 = \frac{4m_1 - m_2}{16m_1 + m_2} g = \frac{(10)(9.81)}{190} = 0.52 \text{ m/s}^2.$$

Since this number is positive, the assumed direction of the acceleration is correct. The 30-kg mass moves upward.

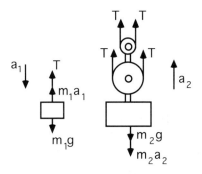

3.21. When a boat goes across an aqueduct, the load does not change. The boat weighs the same as the water it displaces.

3.22. When some of the cargo falls off a ship in a closed canal lock and sinks, it *lowers* the water level. When the cargo was on the ship, it displaced a volume of water that weighed the same as the cargo. When *in* the water, the cargo displaces a lesser volume of water, namely its own volume.

3.23. When a man walks from one end of a lightweight boat to the other, the center of gravity of the man-and-boat system remains fixed because no external force acts on it. So when the man moves one way, the boat moves the other. When the man stops, the boat stops.

3.24. The float swings to the left. When the jar is accelerated, the float swings in the direction of acceleration, not away from it. The reason is that the water pressure behind the float is greater than the pressure ahead of it, great enough to accelerate the float faster than the jar.

3.25. When the sealed glass ball is moved halfway up in the pool, it stays put because its volume and weight remain constant. When the ball with the hole near its bottom is moved up, it experiences lower pressure. The air in the ball pushes out some of the water (that entered when the ball was sunk). The ball loses weight and therefore rises. When the balloon is moved up, it expands, displaces more water, and therefore also rises.

3.26. When a person of mass $m$ at the equator starts to float up in the air, the centrifugal force on him or her, $mr\omega^2$, equals the force of gravity, $mg$, where $r$ is the radius of the earth and $\omega$ is its angular velocity. Then

$$\frac{1}{\omega} = \sqrt{\frac{r}{g}} \text{ seconds per radian}$$

or $\quad \dfrac{1}{\omega} = \dfrac{2\pi}{3600} \sqrt{\dfrac{6.379 \times 10^6}{9.78}}$
$\qquad = 1.41$ hour per rev.

So that fateful day would be 1.41 hours long. This is very nearly the length of the day observed by an astronaut in a low orbit around the earth. Like those people at the equator, he is also "floating in the air."

3.27. If $R$ is the resistance between A and B, the network can be drawn like this:

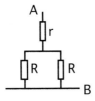

So

$$R = r + \frac{R}{2}$$

or

$$R = 2r.$$

3.28. If $R$ is the radius of the earth in meters, the circumference of the completed pipeline will be $2\pi R+5$ meters, and its radius, $R+(5/2\pi)$. So the height of the supports must be $5/2\pi$ meters or 80 cm. Think of all that work to compensate for a five-meter mistake!

3.29. The portion of the network to the right of C and D is the same as the original network with every resistance doubled. So the original network can be represented by the next figure. If $R$ is the resistance between A and B,

$$R = 4 + \frac{4R}{2 + 2R}$$

and

$$R = \frac{5 + \sqrt{41}}{2} = 5.70 \text{ ohms.}$$

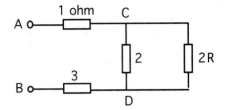

3.30. When the network is terminated in its characteristic resistance $R_o$ as in the figure below, the resistance looking down the line from each section is the same, namely $R_o$. This resistance between E and F is $R$ in parallel with $2R+R_o$. Thus

$$R_0 = \frac{R(2R+R_0)}{3R+R_0}$$

and

$$R_0 = (\sqrt{3} - 1)R.$$

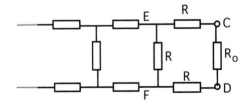

3.31. Open the network at points A and B. The resistance $R$ looking to the right from A and B is the same as the resistance looking to the right from C and D. So the right half of the network can be represented as shown in the figure below. Thus

$$R = 1 + \frac{R+1}{R+2} = \sqrt{3} \text{ ohms.}$$

With the left half of the network included, the resistance between A and B is

$\sqrt{3}/2$ or 0.866 ohms.

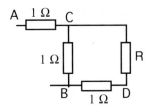

3.32. When $x_f$ is at its critical value, the knife is just about to tip up from the plate, so the sum of moments on the knife about point A is zero, and

$$F x_f = W_k x_k.$$

At the same time the plate and knife are about to tip up as a unit, so the sum of moments on the plate and knife about point B is zero, and

$$F(x_f + a) = W_p b + W_k(x_k - a).$$

Manipulation of these two equations

shows that

$$x_f = \frac{x_k}{\frac{W_p}{W_k}\frac{b}{a} - 1}.$$

You can use this formula to make sure the plate and knife are of sizes that will allow the stunt to work.

3.33. Since the vertical component of the tension in arm A is $W$, the total tension $F_A$ is $W/\sin\Theta$. Thus

$$\frac{W}{\sin\Theta} = 2W$$

and

$$\Theta = 30 \text{ degrees}.$$

The compression in arm B is $F_A \cos\Theta$ or 1.732 $W$.

3.34. When the terminals of the boxes are open-circuited, no current flows in box A, but the resistor in box B draws 10 amperes and dissipates 100 watts. Box B is therefore warmer than box A. If you short the terminals of each box, the resistor in box A will draw 10 amperes and dissipate 100 watts. The resistor in box B will be shorted, draw no current, and dissipate no power. Then box A will

**Solutions to the Problems**

warm up and box B will cool off.

3.35. If the distance between the centers of the two spheres of mass $m$ is $r$, the gravitational force on each sphere is

$$f = \frac{km^2}{r^2},$$

where $k$ is the universal gravitational constant, $6.672 \times 10^{-11}$ m$^3$s$^{-2}$kg$^{-1}$. When the distance $r$ has decreased to $x$ from its original value of $x_0$, the loss of potential energy of the two spheres is

$$-\int_{x_0}^{x} \frac{km^2}{r^2} dr = \left. \frac{km^2}{r} \right|_{x_0}^{x} = km^2 \left[ \frac{1}{x} - \frac{1}{x_0} \right].$$

Since each sphere accelerates to the velocity $(dx/dt)/2$ relative to its original position, its kinetic energy is

$$\frac{m}{2} \left[ \frac{1}{2} \frac{dx}{dt} \right]^2.$$

The kinetic energy of both spheres equals their loss of potential energy:

$$\frac{m}{4} \left[ \frac{dx}{dt} \right]^2 = km^2 \left[ \frac{1}{x} - \frac{1}{x_0} \right].$$

Since the spheres are approaching each other, $dx/dt$ is negative:

$$\frac{dx}{dt} = -\sqrt{\frac{4km}{x_0}} \cdot \sqrt{\frac{x_0-x}{x}}.$$

Let

$$C = \sqrt{\frac{x_0}{4km}}.$$

Then

$$dt = -C\sqrt{\frac{x}{x_0-x}}\, dx.$$

The time required for the distance between the centers of the spheres to decrease from $x_0$ (1 m) to $x_1$ (0.1 m) is

$$t = -C\int_{x_0}^{x_1} \sqrt{\frac{x}{x_0-x}}\, dx$$

or

$$t = C\left[\sqrt{x_0 x - x^2} + x_0 \sin^{-1}\sqrt{\frac{x_0-x}{x_0}}\right]_{x_0}^{x_1}$$

$$= C\left[\sqrt{x_0 x_1 - x_1^2} + x_0 \sin^{-1}\sqrt{\frac{x_0-x_0}{x_0}}\right]$$

$$= 94{,}821.38 \text{ seconds or } 26.34 \text{ hours.}$$

3.36. Construct a terminal at infinity by shorting all of the leads around the periphery of the network. Then connect one of the nodes, like node A in the next figure, to infinity through a current generator that produces one ampere flowing toward the node. Each of the four resistors connected to node A then carries 1/4 ampere away from the node. Now turn off the current generator, and connect the adjacent node B to infinity through another current generator causing one ampere to flow away from that node. Each of the four resistors connected to node B carries 1/4 ampere toward the node. Then when *both* current generators are on, a current of one ampere flows into node A and away from node B. The current in the resistor between the two adjacent nodes is 1/2 ampere, and the voltage drop across the resistor is 1/2 volt. The resistance between the two nodes is therefore 1/2 ohm.

**Little Engineering Problems**

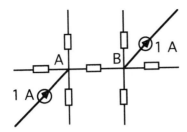

3.37. Let  $A$ = senior
 $B$ = student member
 $C$ = active in branch functions
 $D$ = not on academic probation

The four statements in the newsletter can be written as

1. $\bar{A} B \bar{C} D$

2. $A B D$

3. $A B \bar{C} D$

4. $\bar{A} B C D$

The Karnaugh map following shows that the statements simplify to $BD$. A candidate must be a student member of IEEE who is not on academic probation.

|     | 00 | 01 | 11 | 10 |
|-----|----|----|----|----|
| 00  |    |    |    |    |
| 01  |    | 1  | 1  |    |
| 11  |    | 1  | 1  |    |
| 10  |    |    |    |    |

3.38. You could find the internal resistance of a battery by dividing its open-circuit voltage by its short-circuit current. But measuring the latter would discharge the battery in a hurry. A better way would be to measure the open-circuit voltage and then connect a known resistance $R_1$ across the battery's terminals as in the following figure. Then measure the voltage $V_1$ across the resistor. The internal resistance of the battery is

$$R = \left[\frac{V}{V_1} - 1\right] R_1 .$$

If you had a potentiometer to use as $R_1$, you could adjust it to make $V_1$ half of $V$. Then $R_1$ would equal $R$, and you could read the internal resistance directly on the potentiometer.

If you had an ammeter, you could measure the current $I$ through $R_1$. Then

$$R = \frac{V}{I} - R_1 ,$$

# Little Engineering Problems

but why use two instruments when you need only a voltmeter?

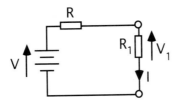

3.39. It didn't take you long to find that $R = 45$ ohms, right?

3.40. The object is a wedge with a notch in its long face:

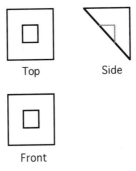

3.41. The object is a cylinder sharpened to an edge by two plane cuts, like the head of a cold chisel. The curve in the side view, shown here, is half of an ellipse.

**Solutions to the Problems**

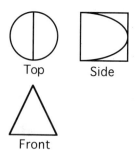

3.42. The connections below aren't as short as possible, but they don't cross each other.

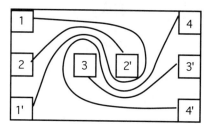

3.43. The signal is, "To be or not to be!"

3.44. Since the specific gravity of the cork ball is 0.2, its weight is

$$\frac{(0.2)(1000)(4\pi)}{3} = 838 \text{ kg} .$$

Handles won't help. We need a fork truck.

3.45. If $X(\omega)$, $Y(\omega)$, and $H(\omega)$ are the Fourier transforms of the input, output, and impulse response of a linear, time-invariant system,

$$Y(\omega) = X(\omega) \, H(\omega) \; .$$

If a frequency $\omega$ is not present in $X(\omega)$, then for that frequency $X(\omega)$ is zero, and $Y(\omega)$ is zero also.

3.46. The earth's rotation gives the target a tangential velocity of $\omega R$, where $\omega$ is the angular velocity of the earth, 1 rev/day, and $R$ is its radius. The tangential velocity of the top of the tower is $\omega(R+h)$, where $h$ is the height of the tower, 100 m. The difference in the two tangential velocities is $\omega h$. If air resistance is neglected, the time required for the weight to fall to the ground is

$$\sqrt{2h/g} \; ,$$

where $g$ is the acceleration due to gravity, 9.78 m/s$^2$ at the equator. So the weight misses the bull's-eye by the distance

$$\omega h \sqrt{2h/g} \; .$$

When the numbers are inserted in the right units, we see that the weight lands 3.3 cm east of the center of the bull's-eye.

3.47. The free-body diagram of one of the cylinders is shown on page 129. The gravitational force $mg \sin A$ pulling the cylinder down the ramp is opposed by the inertial force $ma$, where $a$ is the cylinder's acceleration, and by the force $F$ that gives the cylinder its angular acceleration $\alpha$. Since $a = R\alpha$ and $FR = J\alpha$ where $J$ is the polar moment of inertia of the cylinder,

$$F = \frac{J\alpha}{R} = \frac{Ja}{R^2}.$$

Equating the forces on the cylinder parallel to the ramp, we get

$$mg \sin A = ma + \frac{J a}{R^2} = a \left(m + \frac{J}{R^2}\right)$$

and

$$a = \frac{mg \sin A}{m + \frac{J}{R^2}}.$$

For the wooden cylinder, $J = mR^2/2$, and

$$a = \frac{2}{3} g \sin A = \left(\frac{2}{3}\right)(9.81)(0.5) = 3.27 \text{ m/s}^2.$$

For the steel cylinder, $J$ is very nearly $mR^2$, and

$$a = \frac{1}{2} g \sin A = 2.45 \text{ m/s}^2.$$

The steel cylinder, having the higher moment of inertia, accelerates slower and drops farther behind the wooden cylinder. Notice that the accelerations of the cylinders are independent of their masses or radii.

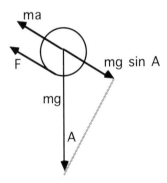

3.48. The figure below shows that the gravitational force pulling the box forward is

$$F_1 = M g \sin A. \tag{1}$$

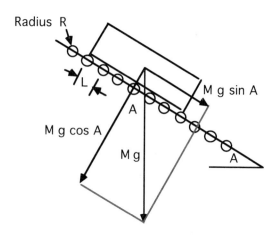

When the box has reached its steady speed, only $m$ of the $n$ rollers under the box exert frictional force on it. The total frictional force holding the box back is therefore

$$F_2 = Mgf\,\frac{m}{n}\cos A. \qquad (2)$$

Equating these two forces produces the result

$$m = \frac{n}{f}\tan A. \qquad (3)$$

The distance the box moves across a roller before the roller gets up to speed is $mL$. Since the box is moving at the velocity $v$, the time required for a roller to get up to speed is

$$t = \frac{mL}{v}. \tag{4}$$

The torque accelerating each roller is

$$T = RF_2 = RMgf\,\frac{m}{n}\cos A. \tag{5}$$

If $a$ and $\alpha$ are the linear and angular accelerations of a sliding roller,

$$T = J\alpha = \frac{Ja}{R}. \tag{6}$$

Combining the last two equations shows that

$$a = \frac{R^2 Mgfm\cos A}{nJ}. \tag{7}$$

Then

$$v = at = \frac{R^2 Mgfm\cos A}{nJ}\cdot\frac{mL}{v}. \tag{8}$$

A combination of Eqs. (8) and (3) is

$$v = \left[\frac{R^2 MgLn\sin A\tan A}{fJ}\right]^{1/2}. \tag{9}$$

With the numbers inserted, the velocity of the box becomes

$$v = 1.44 \text{ m/s}. \tag{10}$$

Eq. (3) shows that when $m = n$, the ramp angle is

$$A = \tan^{-1} \frac{mf}{n} = \tan^{-1}(0.3) = 16.7 \text{ degrees} . \quad (11)$$

If the ramp is any steeper, none of the rollers under the box gets up to speed. The box slides over all of them and accelerates continuously.

3.49. In the side view of the drum shown below, $m$ is the mass of the drum and its contents, and $a$ is their acceleration. The inertial force $ma$ and the gravitational force $mg$ act through the drum's center of gravity. The drum will tip over when the sum of moments of these two forces about the edge of the drum's bottom is zero. Then $25\ mg = 50\ ma$ and

$$a = g/2 = 4.90 \text{ m/s}^2 .$$

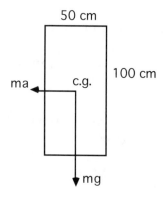

3.50. The wave crests are just a moving pattern, not moving water. A pattern, like the dark spots that run along the line of incandescent lights on an advertising sign, can move at any speed. But real things, like water, radio waves, and surfers, are limited to the speed of light.

# Acknowledgments

This book contains 186 problems, all of which have appeared in the Gamesman section of *Potentials*. Prof. John Kaczorowski of Northeastern University was the first Gamesman editor. Twenty-seven of his problems are included. Readers of the magazine get the credit for submitting 100 of the problems. The current Gamesman editor, the author of this book, contributed the remaining 59 problems. The individual contributions of Prof. Kaczorowski and the readers are listed here with gratitude.

## CHAPTER 1    PROBLEMS REQUIRING ONLY LOGIC

| | |
|---|---|
| 1.1. | Carey Rappaport, MIT, Cambridge, MA |
| 1.2. | David W. Hahn, Baton Rouge, LA |
| 1.3. | Rajiv Gupta, State University of New York, Stony Brook |
| 1.4. | John A. Zizzo, Jr., Staten Island, NY and Amjad A. Soomro, Dhahran, Saudi Arabia |
| 1.6. | Manoj Mehta, Kew Gardens, NY |
| 1.7. and 1.9. | Prof. John Kaczorowski, Northeastern University |
| 1.13. | Hitendrakumar Bhakta, Corpus Christi, TX |
| 1.15. | Vijay Ramaiah, Pratt Institute, and Sundar Sreenivasachar, University of Victoria, Canada |
| 1.17. | John C. V. Ferguson, Florida Institute of Technology |

| | | |
|---|---|---|
| | 1.18. | Ananth Vaidyanathan, Kolhapur, India |
| | 1.22. | Yang Wang, Case Western Reserve University |
| | 1.23. | Dr. Ann Miller, Motorola, Inc. |
| | 1.24. | Lynn Roberts Bengtson, Branford, CT |
| | 1.25. | Charlie A. Jones, Blacksburg, VA |
| | 1.26. | Kristin Kerr, University of Southern Maine |
| | 1.27. | Sally Mack, Hopkins School, New Haven, CT |
| | 1.28. | Manikarnika Ramanujan, College Park, MD |
| | 1.30. | Earl E. C. Angus, Milpitas, CA |
| | 1.31. | David Ng, University of British Columbia |
| | 1.32. | Marcelo Gennari, National University of Tucuman, Argentina |
| | 1.33. | R. Naga V. Sitaram, Cochin University of Science and Technology, India |
| 1.34. to 1.47. | | Prof. John Kaczorowski, Northeastern University |
| 1.48. to 1.65. | | Walter Lau, California State University, Chico |
| 1.66. to 1.76. | | Robert Manning, New Haven, CT |
| | 1.77. | Ajit V. Bhumkar, New Bombay, Maharashtra, India |
| 1.78. and 1.79. | | Prof. John Kaczorowski, Northeastern University |
| | 1.80. | Masoud Rafati, Tehran, Iran |
| | 1.81. | Nho Hong Vo, Reseda, CA |
| | 1.84. | Marius du Plessis, Stellenbosch, South Africa |
| | 1.85. | Ioannis Argyropoulos, Evanston, IL |

## CHAPTER 2  PROBLEMS REQUIRING SOME MATHEMATICS

2.1. James C. Rautio, Syracuse University

## Acknowledgments

2.2. Bruce J. Layman, West Richland, WA
2.3. Eric Herz, IEEE General Manager
2.4. Steve Vernon, University of Southern California
2.5. Prof. John Kaczorowski, Northeastern University
2.6. Eric Herz, IEEE General Manager
2.12. Tom Tsai, Irvine, CA
2.13. Terrance A. Lawrence, Christopher Newport College, Virginia
2.14. Alex Pham, Santa Clara University
2.15. Rajesh Vaswani, Bombay, India
2.16. Terry Glaze, Great Falls, MT
2.18. A. Sriharan, University of South Alabama and Hamilton A. Quant, Silver Springs, MD
2.19. Khalid Zia, New Jersey Institute of Technology
2.21. Paul Griffin, Bryan, TX
2.23. David A. Agront, University of Puerto Rico
2.29. Ronald Fernandes, Varanasi, India
2.31. David Skoll, St. John's, Newfoundland
2.32. Tariq Jamil, Haripur, Pakistan
2.33. M. Ali Amanullah, Karachi, Pakistan
2.34. Ronald Fernandes, Varanasi, India
2.35. Michael Heiss, Vienna, Austria
2.36. Ravindra Chandrashekhar, Fort Worth, TX
2.37. Benton Lai, University of Toronto
2.38. Thavorak Nuth, University of Texas, Dallas and Subramaniam Krishnan, Fresno, CA
2.39. Vijay Gehlot, University of Pennsylvania
2.40. Sudeepto Roy, Baroda, Gujarat, India
2.42. Paul Fargen and Tom Ryan, University of Louisville
2.43. Robert M. Briber, Albany, NY

# Acknowledgments

- 2.44. Nasser Najmi, Esfahan, Iran
- 2.47. Shahriar Sadighi, Tehran, Iran
- 2.49. Ioannis Argyropoulos, Evanston, IL

## CHAPTER 3  LITTLE ENGINEERING PROBLEMS

- 3.8. Cindy Furse, University of Utah
- 3.13. Prof. Jerry Russell, Polytechnic University
- 3.14. Prof. John Kaczorowski, Northeastern University
- 3.15. Tjundewo Lawu, Satya Wacana Christian University, Indonesia
- 3.16. Dick Vega, San Diego State University
- 3.17. Krishnamoorthy, Rensselaer Polytechnic Institute
- 3.18. A. Sriharan, University of South Alabama
- 3.22. Prof. John Kaczorowski, Northeastern University
- 3.27. Subhasis Chaudhuri, University of California, San Diego
- 3.28. Rob Fiore, Hamden, CT
- 3.29. Shankar Hemmady, University of Iowa
- 3.30. Jomi Sebastian, Pondicherry Engineering College, India
- 3.31. Shankar Hemmady, University of Iowa
- 3.32. Michael Heiss, Vienna, Austria
- 3.34. Craig Borden, Silverado, CA and Prof. Gary A. Ybarra, North Carolina State University
- 3.36. to 3.39. Prof. John Kaczorowski, Northeastern University
- 3.41. to 3.43. Prof. John Kaczorowski, Northeastern University
- 3.45. Keivan Bahazi, Sharif University of Technology, Iran
- 3.50. Michael Heiss, Vienna, Austria

Do YOU have an original brainbuster? If so, we'd like to see it! We'll consider your puzzle for publication in future editions of this book.

You can send us your puzzle by printing or typing it, either in the space below or on a separate sheet, and mailing it to:

Brainbusters
c/o IEEE PRESS
PO Box 1331
445 Hoes Lane
Piscataway, NJ 08855-1331

And don't forget to include the solution!

---

Yes! I have an IQ-busting puzzle for possible use in future editions of the *Unofficial IEEE Brainbuster Gamebook*.

Puzzle: _____

_____

_____

_____

_____

_____

_____

_____

_____

Solution: _____

_____

_____

_____

_____

_____

_____

The IEEE PRESS has my permission to reprint this puzzle in future editions of the book if it is found acceptable for publication.

Name: _____  Phone: _____

Affiliation: _____

Address: _____

_____

Signature: _____